Hands-On Networking

From Theory to Practice

Learn the core theory and engage with real-world networking issues with this richly illustrated example-based textbook.

Hands-On Networking provides students with:

- An accurate view of modern networks through detailed coverage of the most relevant networking technologies,
- Carefully designed, affordable laboratory exercises based on open-source software networking tools for hands-on practice with real networking devices,
- Numerous case studies and examples that link theory to practice,
- A bottom-up approach that is easy to follow and perfect for lab-oriented courses.

Maria Luisa Merani is an Associate Professor at the University of Modena and Reggio Emilia, Italy, where she has taught networking courses since 1993. She is an IEEE Senior Member, an Editor of the *IEEE Transactions on Wireless Communications*, and is the author of more than 70 technical papers in the field.

Maurizio Casoni is an Associate Professor in Telecommunications at the University of Modena and Reggio Emilia, Italy.

Walter Cerroni is an Assistant Professor in Telecommunications at the University of Bologna, Italy. His teaching experience covers different aspects of communication networks, with most of the courses integrating theory and laboratory exercises or simulations.

Hands-On Networking

From Theory to Practice

MARIA LUISA MERANI
University of Modena and Reggio Emilia, Italy

MAURIZIO CASONI
University of Modena and Reggio Emilia, Italy

WALTER CERRONI
University of Bologna, Italy

CAMBRIDGE UNIVERSITY PRESS
Cambridge, New York, Melbourne, Madrid, Cape Town, Singapore, São Paulo, Delhi

Cambridge University Press
The Edinburgh Building, Cambridge CB2 8RU, UK

Published in the United States of America by Cambridge University Press, New York

www.cambridge.org
Information on this title: www.cambridge.org/9780521869850

First published 2009

Printed in the United Kingdom at the University Press, Cambridge

A catalog record for this publication is available from the British Library

ISBN 978-0-521-86985-0 hardback

Additional resources for this publication at www.cambridge.org/9780521869850

To Pierangelo and our three enchanting children:
Pietro, Filippo and Margherita.
Maria Luisa

To Francesco, Beatrice and Rita, happiness of my life.
Maurizio

To Lara, for her total love, constant support and extreme patience.
Walter

Contents

Preface   *page* xi

1   Foundations   1

1.1   Signals: time and frequency analysis   1
1.2   A more general notion of bandwidth   5
1.3   Physical media   5
1.4   Network classification   8
       1.4.1   The obvious starting example   8
       1.4.2   Circuit-switched versus packet-switched networks   9
       1.4.3   Distance-based classification   11
       1.4.4   Topology-based and further classifications   12
       1.4.5   Channel-based classification   14
1.5   Transmission options   14
1.6   Network delay   15
1.7   A last miscellanea of concepts   18
       1.7.1   Traffic sources   18
       1.7.2   Service taxonomy   18
       1.7.3   Performance metrics   20
       1.7.4   Congestion and QoS   21
1.8   A few bibliographical notes   22
1.9   Practice: determining the RTT   22
1.10   Exercises   26

2   Architectures and protocols   29

2.1   Who's who in the telecommunication world   29
2.2   OSI Model: the seven-layer approach   31
2.3   TCP/IP protocol suite   37
2.4   IP: Internet protocol   40
       2.4.1   Public and private IP addressing   42
       2.4.2   Classless IP addressing   46
       2.4.3   Subnetting and supernetting   48
       2.4.4   The delivery of IP packets   52
2.5   TCP: transmission-control protocol   54

2.6	UDP: user datagram protocol	58
2.7	Exercises	58

3   Ethernet networks   **61**

3.1	Multiple access	61
	3.1.1 Carrier sense multiple access strategies	65
3.2	IEEE 802.3 and the IEEE 802 project	67
	3.2.1 Reference topologies	68
	3.2.2 MAC sublayer	70
	3.2.3 Physical layer	75
3.3	Twisted-pair cabling standards	80
3.4	Practice: address resolution protocol	87
3.5	Practice: NIC configuration	92
3.6	Practice: a campus network layout	95

4   Wireless networks   **97**

4.1	Wireless LAN	97
	4.1.1 The basket of 802.11 standards	97
	4.1.2 Physical layer evolution	98
	4.1.3 Architecture and MAC basic mechanisms	101
	4.1.4 The need for quality of service and the 802.11e document	107
	4.1.5 IEEE 802.11 frame format	110
	4.1.6 Recent enhancements: the 802.11n document	112
	4.1.7 The Wi-Fi Alliance	113
4.2	Wireless MAN	113
	4.2.1 Physical layer	114
	4.2.2 MAC features	118
	4.2.3 IEEE 802.16 frame format	122
	4.2.4 Scheduling services	123
	4.2.5 WiMAX Forum	124
4.3	WPAN: wireless personal area networks	124
4.4	A glimpse of wireless mesh networks	125
4.5	Practice: capturing 802.11 data and control frames	127
4.6	Practice: inspecting 802.11 management frames	131
4.7	Practice: cracking the 802.11 WPA2-PSK keys, perhaps . . .	132
4.8	Exercises	134

5   LAN devices and virtual LANs   **136**

5.1	Repeaters and bridges	136
5.2	Main features of bridges	138
5.3	Switches	139
5.4	Virtual LAN	139

5.5	Overview: VLAN definition and benefits	140
5.6	VLAN classification	141
5.7	VLAN on a single switch	143
5.8	VLAN on multiple switches	145
	5.8.1 The need for tagging and virtual topology	145
	5.8.2 IEEE 802.1Q frame tagging	147
5.9	Inter-VLAN communications	148
5.10	Practice: switch management and VLAN configuration	151
	5.10.1 Switch management	151
	5.10.2 VLAN configuration	155
	5.10.3 Inter-VLAN communication	157
5.11	Exercises	159

6	**Routers**	**161**
6.1	What is a router?	161
6.2	Functions and architectures	162
6.3	Table look-up implementation	165
6.4	From routers to middleboxes: firewalls and NATs	168
6.5	Practice: routing and forwarding table	173
6.6	Practice: firewalls and packet filtering	176
6.7	Practice: network address translation	182

7	**Routing fundamentals and protocols**	**187**
7.1	Routing algorithms	187
	7.1.1 The Bellman–Ford algorithm	189
	7.1.2 The Dijkstra algorithm	192
7.2	Routing protocols	193
	7.2.1 Distance vector protocols	193
	7.2.2 Link state protocols	194
	7.2.3 Distance vector, path vector or link state?	195
7.3	Routing in the Internet	196
	7.3.1 Routing information protocol	198
	7.3.2 Open shortest path first	200
7.4	Practice: RIP configuration	203
7.5	Practice: OSPF configuration	210

8	**Wide area networks and user access**	**220**
8.1	The xDSL family	220
8.2	The X.25 network	222
8.3	Integrated services digital network	225
8.4	The frame relay service	227
8.5	B-ISDN and ATM	231

8.6	MPLS principles	236
8.7	Practice: MPLS router configuration	239
	8.7.1 Basic LDP configuration	239
	8.7.2 MPLS traffic engineering	244
8.8	Exercises	247

References	248
Index	253

Preface

The topics this book touches lie within the networking field, an exciting area that in the last 20 years has experienced a stunning growth and gained an increasing popularity. Just as previous ages of modern society have been marked by technological advancements that significantly shaped them, from transistors to personal computers, our life is now molded by emails, our work and leisure time clocked by websites, our children daily accompanied by the Internet. What lies behind this boiling surface? What infrastructures and communication rules allow us to connect to the Internet from home through an ADSL connection? How does information travel on a high-speed backbone from our office to an overseas destination? Through a rigorous yet practical approach, the aim of this volume is to provide all the concepts needed for a thorough knowledge of networking technologies, as well as to breed the development of agile skills in modern network design.

After laying the common language foundations and the basic concepts and terminology within the field, the book is committed to a critical treatment of the technologies, protocols and devices adopted in contemporary networks. A special emphasis is placed on building an effective competence in all subject areas: hence, each topic is complemented by guided and commented practices, where proficiencies are challenged through real problems. The aim is to strengthen the abilities needed for present-day network design. The chapter structure reflects the authors' pedagogical view: first build good foundations and gain expertise in each topic, then consolidate and confront real networking issues.

With only a few exceptions, the practical examples described in the book are based on open-source software networking tools. The rationale behind this choice is to give the reader the chance to experiment on real networking devices and tools at relatively limited cost, as most of the practices can be replicated using a few PCs and some very common networking hardware. However, this open and accessible methodology does not limit the technical quality of the examples: the same conceptual approach holds when using commercial – and expensive – networking equipment.

Undoubtedly, this book has been conceived to provide university students with strong competence in the networking field, an ambitious task that requires the blend of two distinct ingredients. First, to master theory and concepts rigorously, acquiring a critical methodology in approaching problems: the latter is the ultimate lesson a university professor should strive to teach! Second, to tame real systems and real problems, a goal where universities sometimes fall short. Our hope is to contribute to partly bridge the gap, pouring our teaching and professional experiences into this writing adventure.

Who are the intended recipients of the volume? Both university students and professionals willing to achieve a solid foundation in the networking field. It might be used as a textbook for a one-semester undergraduate or first graduate course complemented by its laboratory class, both in electrical engineering and computer science departments. For professionals and practitioners in industry, its pragmatic approach allows one to easily frame the concepts and design issues lying behind the interconnecting world. Deliberately, the book does not require any specific competence in the field of probability theory, statistics and stochastic processes.

If the reader is ready to take his or her first swimming lesson, it is time to commence: in the allegory, the water is the networking world, the book is the pool and the goal . . . well, not the Olympics: setting – just slightly – less prestigious goals does not depend on the athlete, rather, on the instructors. Nevertheless, at the end of the entire swimming course the authors of the book will consider a good result to have their pupil swim fearlessly and be in good shape.

So, let us dive in . . .

1 Foundations

This is the chapter where the reader first approaches the world of networks. The preliminary step to take consists in providing a succinct introduction to the analysis of the signals that, either in the electrical or in the optical domain, carry data information across networks.

Also, before throwing onto the floor the first Lego bricks and the instructions to let the reader build the composite view of modern computer networks, a review of the "fabric" of these bricks is needed. A light excursus is thus provided, to gain a basic acquaintance with copper wires, fiber optics and the radio channel, each transmission medium being employed in specific networking contexts.

Network classification comes next: it is a useful exercise to gain familiarity with the main ideas and the terminology peculiar to the world of networks. This is followed by the introduction of the crucial notion of delay, as well as by a miscellanea of concepts that cross different areas, spanning from sources of network traffic to service classification, from performance metrics to quality of service.

Following a pattern that will shape the exposition throughout the entire book, a direct experience is proposed and commented at the end of the chapter, to let the reader confront the real world via a first-hand adventure.

1.1 Signals: time and frequency analysis

Indeed, our aim is to talk about networks. Before doing so, however, we have to recall the fundamentals of signal analysis, to understand what type of information we are moving around.

The more intuitive approach takes us to imagine – digital – information streams traveling among computer facilities and crossing network devices. If we think of the binary transmission of the single hexadecimal value 4F, i.e., byte 0100-1111, then the corresponding signal in the time domain could be represented as in Fig. 1.1, and could, for instance, be a voltage level at the transmitter output.

This simple description provides a way to introduce two parameters pertaining to a digital signal: the bit time T_{bit}, and the bit rate B_r. The first is the time it takes for a bit to be "pushed out" of the transmitting device; the second is its inverse, $B_r = 1/T_{bit}$. The higher the bit rate B_r, the more rapid are the signal fluctuations.

Beyond this intuitive analysis in the time domain, we are also interested in examining the properties of signals in the frequency domain.

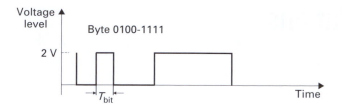

Voltage level

Byte 0100-1111

2 V

T_{bit}

Time

Fig. 1.1   A possible representation of byte 4F via voltage levels

To this end, we recall that, provided some conditions hold, every aperiodic, real signal $x(t)$ defined for $-\infty < t < +\infty$ can be Fourier transformed via the following relation:

$$X(f) = \int_{-\infty}^{+\infty} x(t)e^{-j2\pi ft}\,dt, \tag{1.1}$$

$X(f)$ being the counterpart of $x(t)$ in the frequency domain.

The corresponding inverse transform is

$$x(t) = \int_{-\infty}^{+\infty} X(f)e^{j2\pi ft}\,df. \tag{1.2}$$

As $x(t)$ is a real function, we have that for the complex function $X(f)$ the equality

$$X(f) = X^*(-f) \tag{1.3}$$

holds, where $(\cdot)^*$ denotes the complex conjugate; equivalently,

$$\begin{cases} |X(-f)| = |X(f)| & \text{and} \\ \arg X(-f) = -\arg X(f) \end{cases}, \tag{1.4}$$

so that $x(t)$ can also be written as

$$\begin{aligned} x(t) &= \int_{-\infty}^{+\infty} |X(f)|e^{j\arg X(f)}e^{j2\pi ft}\,df \\ &= \int_{-\infty}^{+\infty} |X(f)|\,[\cos(2\pi ft + \arg X(f)) + j\sin(2\pi ft + \arg X(f))]\,df \\ &= \int_{-\infty}^{+\infty} |X(f)|\cos(2\pi ft + \arg X(f))\,df \\ &= \int_{0}^{\infty} V(f)\cos(2\pi ft - \varphi(f))\,df, \end{aligned} \tag{1.5}$$

where

$$V(f) = 2|X(f)| \tag{1.6}$$

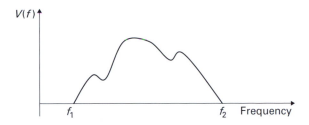

Fig. 1.2   Frequency bandwidth of a signal

and

$$\varphi(f) = -\text{arg}X(f). \tag{1.7}$$

The $V(f)$ function provides the so-called amplitude spectrum for the original signal $x(t)$, and $\varphi = -\text{arg}X(f)$ its phase spectrum.

More importantly for us, via $V(f)$ we introduce the concept of frequency bandwidth of $x(t)$: if $V(f)$ has the behavior that Fig. 1.2 exemplifies, then we affirm that $[f_1, f_2]$ is the frequency bandwidth B_f of signal $x(t)$, whose width is $f_2 - f_1$ hertz (Hz being the abbreviated form).

In this case Eq. (1.5) modifies into

$$x(t) = \int_{f_1}^{f_2} V(f)\cos(2\pi ft - \varphi(f))df , \tag{1.8}$$

which is subject to a nice physical interpretation: within the $[f_1, f_2]$ interval, $x(t)$ appears as the "sum" of infinite cosine terms, the generic term being located at frequency f, with amplitude $V(f)df$ and phase $\varphi(f)$.

Moreover, we observe that, if the function $x(at)$ is considered, $a > 0$, then its Fourier transform is given by

$$\int_{-\infty}^{+\infty} x(at)e^{-j2\pi ft}dt = \frac{1}{a}\int_{-\infty}^{+\infty} x(z)e^{-j2\pi f\frac{z}{a}}dz = \frac{1}{a}X\left(\frac{f}{a}\right). \tag{1.9}$$

This simple result reveals that, when $0 < a < 1$, i.e., when the signal $x(at)$ displays slower variations with respect to $x(t)$, then the low frequency components become predominant in its Fourier transform. Qualitatively, the frequency bandwidth shrinks: if it is B_f for $x(t)$, it is aB_f, $0 < a < 1$, for $x(at)$.

In a corresponding manner, we can immediately conclude that for $a > 1$ the new signal $x(at)$ exhibits more rapid fluctuations in the time domain, and that its frequency spectrum widens with respect to $x(t)$.

Let us now take advantage of this statement when examining the streams of binary digits exchanged over computer networks: recalling the notion of bit rate B_r, we can conclude that, the higher B_r, the wider the frequency bandwidth B_f of the digital signal will be. As our aim is to transmit signals bearing more and more information per unit time, we forcedly have to handle ever increasing frequency bandwidths.

Simple as it sounds, this is definitely a first conclusion to remember.

Step over now, and observe that a digital signal needs to be conveyed through a communication channel, which typically attenuates, delays and, alas, may also distort the input signal.

As a matter of fact, any communication channel is limited in bandwidth to an interval of frequencies where its effects on the conveyed signal are tolerable: typically, this constraint has to be ascribed to the physical limitations of the medium and to the components of the transmitter and the receiver.

If the bandwidth of the input signal $x(t)$, say $[f_1, f_2]$, mainly falls within such interval, what we collect at the channel output is a signal that closely resembles the input in shape. In contrast, if some of the frequency components of $x(t)$ fall outside the channel bandwidth, they will be filtered out, and this will result in a distorted output. If the communication-channel filtering is too drastic, it becomes impossible to recover the information content of the input signal out of the distorted output signal.

It is immediately concluded that, if we aim at transmitting digital signals at high bit rates, we have to employ communication channels featuring wide bandwidths.

In addition to this, we need to observe that transmission can occur either in the native frequency bandwidth of the signal, a circumstance described by the term "baseband transmission," or in a different frequency band, for which the term "bandpass transmission" is adopted. "Baseband" and "bandpass communication channels" also belong to the common glossary of this field.

So, given that a digital signal of the type shown in Fig. 1.3 exhibits a frequency bandwidth which includes frequency $f = 0$ and the nearby (low) frequencies, this signal can be sent over a baseband channel featuring an adequate bandwidth (in this case the channel is acting as a low-pass filter).

It can, however, also happen that the information-bearing signal has to be transmitted over a bandpass communication channel.

How do we handle this mismatch?

We have to require that the *transmitted* signal bandwidth falls within the communication channel bandwidth, otherwise no information delivery process will successfully take place. More explicitly, this calls for a further operation, which roughly corresponds to the translation of the digital signal spectrum to the frequency bandwidth where the channel acts as a bandpass filter: the signal has to *modulate* a suitable carrier frequency.

This is exactly what happens, for example, in radio communications.

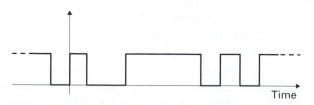

Fig. 1.3   A digital signal in the time domain

1.2 A more general notion of bandwidth

We know by now that a communication channel can support a given, maximum bit rate: it is expressed in bit/s, although Mbit/s and Gbit/s are more appropriate scales for the channels used in computer networks.

As we previously asserted, this rate directly depends on the characteristics of the transmission medium the communication channel is built upon, as well as on the hardware and software characteristics of both transmitter and receiver: confining our attention to the media, it depends on the frequency bandwidth effectively available to deliver the information signal undistorted and not significantly attenuated.

In the networking field, the transmission rate of a communication link and more generally of networks is often referred to as "bandwidth." Internet service providers (ISPs) advertise ADSL connections with 20 Mbit/s downstream bandwidth; sellers of network switches praise their apparatus as offering 1 Gbit/s bandwidth.

Purists shiver, as historically the notion of bandwidth belongs to the frequency domain, and bandwidths are measured in Hertz, not bit/s. True, without any doubt.

There is not even numerical coincidence between the maximum bit rate a channel can sustain and its frequency bandwidth: it is not true that in a communication system built on a channel whose bandwidth is 100 MHz, the maximum achievable transmission rate is exactly 100 Mbit/s (the actual ratio between rate and channel bandwidth can also be significantly greater than one).

What can settle the matter is to use the term *layer-1* (or *physical layer*) bandwidth to tag the old, familiar frequency bandwidth and *layer-2* (or *data-link layer*) bandwidth to identify the nominal transmission rate at which a link or a network carries the bits.

We are aware that the *layer-1*, *layer-2* attributes and their equivalents sound arcane at this point of the book, but in the next chapter they will gain a proper meaning. So, let us agree on their usage and simply observe that these terms inherently refer to the process of information transmission at different abstraction levels.

Talking about layer-1 bandwidth implies a tight attention to the media that carry the information bits; talking about layer-2 bandwidth indicates that the focus shifts onto the information itself.

The next questions we pose are: to what type of physical media do we resort to build the required communication channels? And what are their basic features? The answers come next.

1.3 Physical media

We first introduce those transmission media where the signal travels via guided propagation: among copper wires, where information bits are conveyed to their destination within the electrical domain, we mention twisted pair and coaxial cable; within the optical domain, where the information bits we aim at transferring are represented by either the presence or the absence of a pulse of light, the rails we use to move information around are optical fibers. Next, we will discuss the usage of the radio channel, a transmission medium where the signal, in the form of electromagnetic waves, propagates freely.

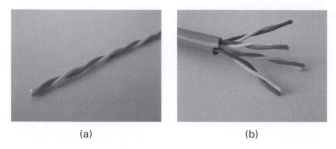

(a)   (b)

Fig. 1.4   A twisted pair and a UTP cable

In telephone networks, twisted pair is by far the most widespread transmission medium: it is employed in the local loop, i.e., in that portion of the network going from the subscriber's premises to the nearest end office of the telephone company. In computer networks, and, more accurately, in local area networks (LANs), twisted pair is also very popular. Honestly though, it bears different characteristics with respect to its telephone network cousin: overall, it is of better quality, a feature not that hard to understand even at this point of the chapter, given we can imagine that the intent is to squeeze more out of a LAN wiring than out of a modest telephone line, whose wires display a bandwidth of a few hundreds of kilohertz (kHz).

What does a twisted pair look like? An example is shown in Fig. 1.4(a), which helps understand where the name comes from: two copper wires, whose diameter is of milli-metrical order, are gently twisted together; for the sake of cabling, several of these pairs are then grouped within a same cable, as Fig.1.4(b) illustrates: note that each wire is insulated and that there is a protective sheath that wraps all the four pairs of this example. Having four pairs is not an accident: we will see it is exactly the number adopted by LAN cables. The term for this solution is unshielded twisted pair (UTP) cable: it is flexible and therefore easy to install, and also cheap. We will meet it again in Chapter 3, where we will discuss in detail its usage and its limits. For now we simply observe that good quality twisted pair displays a frequency bandwidth of a few megahertz; its usage is, however, confined to relatively short distances, a few hundred meters.

Coaxial cable has a different role with respect to twisted pair: no longer used in LANs, broadband coaxial cable originally survived in the separate field of analog – not digital – broadcasting of television channels. It then gained renewed interest for computer networks when cable TV companies began providing Internet connectivity: their subscribers, equipped with appropriate devices called cable modems, can transmit data on broadband coaxial cable at considerable rates.

A coaxial cable is shown in Fig. 1.5. Here too we find two copper conductors: the inner is a cylindrical core, surrounded by a dielectric insulator; the outer is a metallic shield, protected by a plastic jacket. Coaxial cable can be moderately bent and twisted without altering its main characteristics, the reason being that the electromagnetic field is confined between the inner conductor and the shield. In terms of quality, broadband coaxial cable does not shine, but it can be employed for frequencies up to several hundreds of megahertz and for distances up to one hundred kilometers.

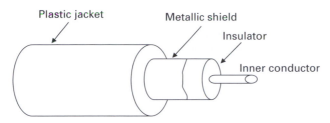

Fig. 1.5   Example of a coaxial cable

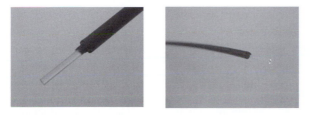

Fig. 1.6   An optical fiber

Among the guided media, optical fiber is by far the best performing choice: its geometry shows the presence of a glass core, surrounded by a glass cladding with a different propagation index and coated by a few protecting layers: some illustrative pictures of a fiber appear in Fig. 1.6. Transmission over fiber occurs in the infrared region, in three distinct wavelength bands, centered at 850 nm, 1300 nm and 1550 nm. Two main fiber categories exist: single-mode and multi-mode fibers. In the former, light pulses propagate within the core in single mode, i.e., with no reflections against the cladding; single-mode fibers commonly display a core diameter of 8 to 10 μm, and operate in the first two bands. Multi-mode fibers, on the contrary, owe their name to the multitude of paths that the light takes along the core. They operate in the second and third wavelength bands.

The attenuation that the fiber introduces is extremely low; its bandwidth is huge, as compared to the competing media, and allows very high data rates, of the order of tens of Gbit/s. Photonic devices for signal generation and detection are currently highly reliable. Moreover, fiber is lightweight, of small size, immune to electromagnetic interference, hard to tap and does not corrode. No surprise that it is the preferred transmission medium in several networking contexts!

As regards the radio channel, here the transmission medium is the ether: the signal bearing the desired information propagates freely, no sharp boundaries delimit the area where it can be heard. This is the strength of communications that rely upon such physical medium: users of a wireless LAN can readily move their laptop, free from the hassle of wiring; subscribers of UMTS mobile radio networks can experience the "always on, always connected" promise, having Internet access in places not reached by cables (we should honestly add "almost" every time and every place).

The radio frequencies employed by most wireless LANs and mobile radio systems fall within the ultra-high-frequency range (UHF), spanning from 300 MHz to 3 GHz.

The bad news is that this portion of radio spectrum is fairly crowded and its usage often requires a license; moreover, the UHF wireless channel is a tough medium, where transmissions are impaired by strong attenuation, multipath propagation, interference from simultaneous transmissions, all statistical in nature. To make things even worse, wireless users move around in a hardly predictable manner, often at high speed, so that the channel characteristics vary both in frequency and time. It is definitely not that trivial to guarantee both high transmission rates and excellent quality on a radio channel!

Finally, it is not always an advantage potentially to have anyone hear anyone else's communication: careful mechanisms have to be put in place to secure the privacy and integrity of the messages exchanged in wireless settings.

1.4   Network classification

1.4.1   The obvious starting example

Having declared what are the transmission media that allow to build the physical infrastructure of networks, let us jump to the most popular and perhaps misused example of networks: the Internet. We define it as a "network of networks" sharing one attribute: all devices over the Internet speak a common language, known as the TCP/IP protocol suite. Its discussion deserves several pages, and we have chosen the next chapter to illustrate it. For now it is important to concentrate on the salient features of the Internet infrastructure, because its heterogeneity and constructive mess already contains all the elements we need to classify networks.

Fig. 1.7 provides a first, necessarily simplified view: the intent is to portray the myriad of access networks constituting the edges of the Internet, interconnected via a tiered

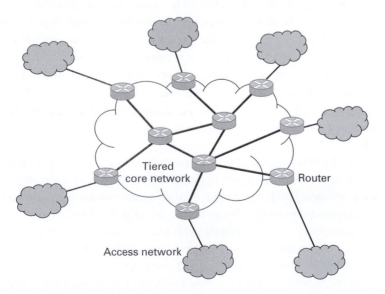

Fig. 1.7   A simplified view of the Internet

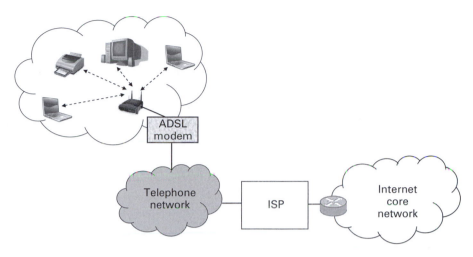

Fig. 1.8   A residential wireless LAN connected to the Internet

core network. Network devices called routers provide the desired connectivity between different domains.

Fig. 1.8 sketches a more familiar portion of the whole picture: in this instance a residential customer is equipped with a small wireless LAN, connected to the Internet through its telephone line and an ADSL modem: after a few hops covered via the telephone network, we meet the router of the ISP guaranteeing the user Internet access.

Routers: what are their main tasks? For the time being, it suffices to know that these nodes are the busy Internet servants: their role is to connect its different constituent networks and to make the information travel from source to destination, properly sized in chunks generically termed *packets*.

1.4.2   Circuit-switched versus packet-switched networks

The word packet immediately gives us a way to distinguish between two classes of networks, which radically differ in the way resources are shared among users: the alternatives are *circuit-switched* and *packet-switched* networks. The Internet is a packet-switched network.

Let us comment on circuit-switching first, according to the same chronological order that the introduction and deployment of these technologies followed. In circuit-switched networks, before any communication can take place, it is mandatory to determine a path between source and destination, and statically allocate resources for that communication along the entire path, i.e., build the circuit. Only after the resource assignment procedure has been successfully completed can the communication take place. Resources are released – the circuit is torn down – as soon as the communication process is over. If for any reason a resource is missing along the path, then the communication cannot occur.

In the past, the telephone network, often referred to as the public switched telephone network (PSTN), was the par-excellence example of a circuit-switched network. To set up a circuit in it means to dedicate resources exclusively to the call within the network switching centers, as well as to assign the call some capacity on the traversed telephone trunks. This hides a significant and time-consuming signaling effort: the circuit set-up request has to propagate along the entire path to the called party, and once successful, a call-accept signal has to find its way back to the calling party (as a result, a ringing tone is sent to the calling telephone, while the called telephone rings).

Currently, the way the PSTN operates is much more articulated: the network backbone of telephone carrier operators also resorts to packet-switching to carry voice calls, but we will not deepen this point.

The approach is completely different in packet-switched networks, where from the very beginning the end users willing to communicate were only computers: the name recalls that the information crosses the networks split in packets. At this point of the chapter, we simply state that the packet format encompasses a field for data bits and one or more fields bearing control bits.

Two different categories exist within the packet-switched family: virtual circuit networks and datagram networks. The Internet core is a datagram network.

In virtual circuit networks, a "virtual circuit" (we could read a "route") is built between source and destination before packets of the information stream can start traveling along that circuit. This concise description might suggest a similarity to circuit switching: it is not so, there is a strong conceptual novelty here, that we now explain. To set up a virtual circuit means to draw a route across the network that all packets of the virtual circuit will follow, but it does not imply that resources are deterministically reserved in advance. Instead, when packets belonging to a virtual circuit travel along the network, they *dynamically* share the capacity of the traversed links with other packets of different virtual circuits; when they arrive at the routers, where they are typically stored and then forwarded to the desired output link, they *dynamically* share the router resources (the power of the processor the router is equipped with, the input and output buffers of the router).

This *statistical multiplexing* of network resources, as opposed to the *deterministic multiplexing* that circuit-switched networks implement, has an immediate, tangible effect: packets might queue up at routers, incurring random delays, which depend on the amount of load each router has to handle. This is something that cannot happen in circuit-switched networks, where no storing at network nodes is required.

The concept behind the virtual circuit approach is represented in Fig. 1.9: in this simple example, packets belonging to three distinct virtual circuits share resources along various paths pinned through the network.

Finally, datagram technology. This is the simplest and most anarchic solution: every packet (a *datagram*, as it was originally called at the beginning of Internet history) of an information flow is individually routed through the network. No static path is built in advance, so that in principle every packet of the stream can follow a different road toward its final destination. To make this possible, the control field of the packet has to bear the ultimate destination address: each router crossed by the packet reads this address to decide the most appropriate output link to forward the datagram. The packet, therefore, proceeds hop-by-hop toward its final recipient, every

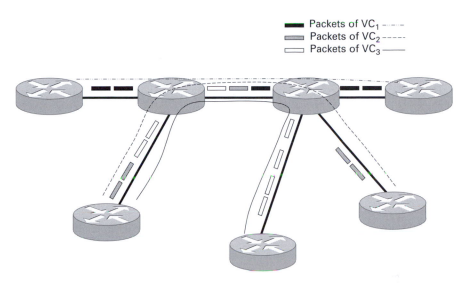

Fig. 1.9   A few virtual circuits traversing a network

crossed router taking it a little bit closer. As in virtual circuit networks, here too data-grams queue up in routers and experience unpredictable delays before reaching the destination.

If a router failure occurs, if some links are close to overloading, it is relatively easy in datagram networks to divert datagrams via alternative routes, provided these are available. This is not so with virtual circuits, which are abruptly terminated if a router they cross experiences an outage condition.

Carrying on the comparison, another significant difference is that datagram networks do not require their routers to maintain state information about the flows of packets that traverse them, so that their operating mode is termed *stateless*. Routers of virtual circuit networks do: they have to know all virtual circuits that currently traverse them, hence, they retain *per-connection-state* information, a burdensome requirement, which poses scalability limits on such devices.

We will pick up again the datagram network topic as we proceed in the exposition: the reader will find thorough details about the tools and the rules that govern the routing process over the Internet in Chapter 2 and also in Chapter 6.

1.4.3   Distance-based classification

Let us proceed in our classification, and distinguish computer networks on the basis of the distance they span. Fig. 1.10 lists the different distances, along with a tangible indication of the location of the network nodes, and the corresponding network names.

On a very limited scale, we encounter personal area networks, PANs: this is the realm of the (not so) futuristic wearable networks, where the connected devices are really close.

Next come local area networks, LANs: they are found in enterprises and university campuses, often interconnect hundreds of nodes and cover areas whose diameter is of the

Distance	Location	Network name
1 m	Body, room	PAN
1 km	Campus	LAN
10 km	City	MAN
100 km	One or more countries	WAN

Fig. 1.10   Network classification by distance

order of a kilometer. Residential LANs are also frequent nowadays: typically, a laptop, mom's old PC, the children's new PC and a printer all belong to the same LAN, often wireless.

Metropolitan area networks, MANs, cover wider areas, usually comparable to the size of a town. Wide area networks, WANs, complete the picture, as they span an entire country or even continents.

Recalling the starting example of the Internet, the access networks at its edges are usually LANs, whereas its backbone is made up of several WANs; MANs can interconnect the two.

1.4.4   Topology-based and further classifications

Networks can also be classified on the basis of their topology; it comes naturally to represent them resorting to graphs, where the vertices are the network nodes and the edges are the physical links between the nodes.

The elementary network topologies are reported in Figs. 1.11 and 1.12: in the first we can easily spot the bus, the star, the ring and the tree topologies; in the second, two different mesh topologies are shown. In a mesh network there are at least two pathways to each node; moreover, a mesh network can be fully connected, as in the first case depicted in the second figure, meaning that every node of the network is directly connected to every other node, or partially connected, when some of the connections are missing.

The bus topology was originally very common in LANs, but was then replaced by the star topology, now the reference for this category of networks: nodes communicating via the star center are typically end-user devices, like personal computers, printers and servers. Ring networks are more common in the metropolitan and wide area network domain, in MANs and WANs, whereas trees are not that frequent. Finally, mesh networks are, to a great extent, the reference for WANs: here the vertices of the graph do not represent end users, but routers.

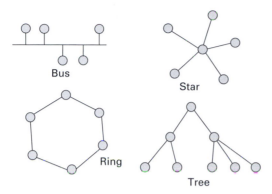

Fig. 1.11   Elementary network topologies

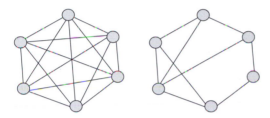

Fig. 1.12   Mesh topologies

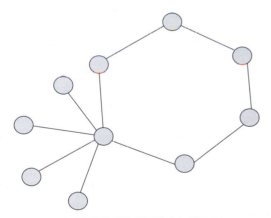

Fig. 1.13   An example of a mixed topology

Stepping back to the Internet examples, we notice that the core network of Fig. 1.7 is a mesh; the wireless LAN of Fig. 1.8 has a star topology, its access point being the star's center.

Networks obeying these elementary topologies can be combined, to create more articulated layouts, as Fig. 1.13 exemplifies, where a mixed ring and star topology is reported.

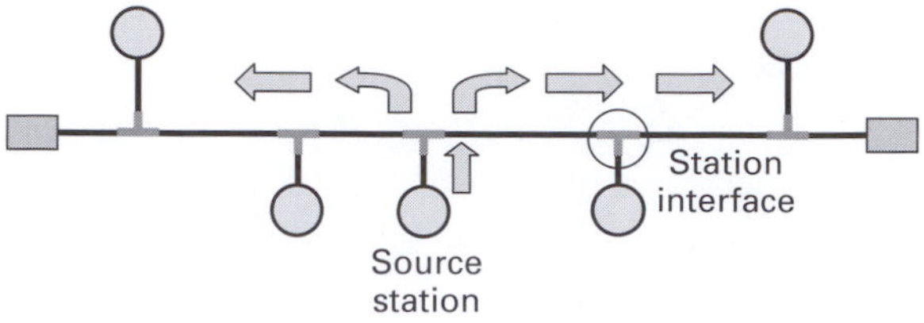

Fig. 1.14   A network built on a wired bus

Further metrics can, of course, be introduced and additional distinctions among networks listed: access networks and core networks have already been cited, as well as wired and wireless networks.

New criteria can also emerge from the most recent settings where networks are spreading: the increased usage of sensor networks, connecting together several, often a multitude of specialized transducers, suggests discrimination by the number of interconnected devices. Delay-tolerant networks, i.e., networks required to operate effectively in environments characterized by extremely high latencies, advocate the introduction of delay as an additional metric to classify different solutions.

The networking field is a highly dynamic world: it is no surprise that new terminology is coined and novel logical schemes are introduced even in the conceptually simple scenario of network classification.

1.4.5   Channel-based classification

We mention here an additional possibility for classifying networks, based on the characteristics of the communication channels the network is built on.

This can be broadcast in nature, meaning that "everyone connected to it can hear everything." A typical example is a wireless LAN: all stations receive the packets transmitted by the central access point over the common radio channel (although only the intended receiver is able to process and interpret their content). A LAN built on a wired bus represents another example: the bus is the broadcast channel, and packets transmitted over it by one station reach the interfaces of all other stations connected to the bus, as Fig. 1.14 indicates.

In contrast to broadcast communication channels, we can have point-to-point communication links, which, differently interconnected, give rise to various networks. The Internet core of Fig. 1.7 is an example of usage of point-to-point links.

1.5   Transmission options

Now that we have declared what kind of media, what technologies and what topologies are available to build networks, we devote a few words to the introduction of some recurring concepts in the networking field.

First of all, transmission types: these are classified as unicast, broadcast and multicast.

A *unicast transmission* implies a one-to-one flow of data from the source to a single receiver.

In reverse, a *broadcast transmission* entails the source transmitting to the totality of the receivers. These could be all users of a LAN: indeed, broadcast transmissions are easily performed over the topologies that local area networks display, the bus and the star.

Somewhat in between, there lies *multicast transmission*, where the source simultaneously transmits to a subset of the receivers.

Next, it is useful to explain what the terms simplex, half-duplex and full-duplex transmission mean; for the sake of simplicity, the focus is on the exchange of information between two nodes in a network.

A communication is defined *simplex* when data flow in one direction only.

If the communication is bidirectional, but data cannot simultaneously travel in both directions between the nodes, *half-duplex* is the term to employ.

If the data exchange can occur in both directions at the same time, the communication is *full-duplex*.

Time division duplex (TDD) is the technique that allows bidirectional communication over a single communication channel, in a half-duplex manner: each node can transmit only in its assigned time intervals, which do not overlap with the intervals the second node employs to transmit in the opposite direction.

A few, intuitive solutions to support full-duplex transmissions can also be cited: the most obvious is to employ separate media, i.e., separate physical communication channels between the two nodes. On a single physical medium, a conceptually simple approach that warrants full-duplex communication is frequency division duplex (FDD): this technique divides the frequency (layer-1) bandwidth of the channel into two sub-bands, each exclusively assigned to the communication in one direction.

1.6   Network delay

Packets traversing a computer network, be it a LAN, a WAN, or any combination of different networking solutions, will experience some delay.

Let us inspect the different terms that contribute to it and, to understand them, let us begin with the specific situation of Fig. 1.15: a single point-to-point link between two routers. We name the two interfaces at the ends of the link A and B.

When a packet enters the first router from one of its interfaces and needs to proceed along the A–B link, the router will have to inspect its control field to understand what is the proper interface to forward the packet to: this will introduce some processing delay, not that remarkable really, and we call it T_{proc}.

Fig. 1.15   A single point-to-point link between two routers

Once the outgoing line has been chosen, there is, however, no guarantee that the packet will immediately find it available: the router could be busy transmitting another packet on interface A and some other packets could already be waiting their turn for transmission along that link.

The packet will diligently have to line up in queue with these packets, until it will be picked up and forwarded along the link, via interface *A*.

This second delay is the queueing delay, T_q: it is a random variable, which depends on the rate and pattern at which packets arrive to the routers, as well as on their size; also the queueing discipline that the router adopts to schedule next transmission along the output link plays a role in its determination.

Now imagine that the turn to transmit the packet finally comes.

If the packet is *L* bits long, and the link transmission rate is *C* bit/s, a time interval

$$T_{tx} = L/C \qquad (1.10)$$

will elapse from the time the first bit of the packet appears at interface *A* until the last bit of the packet leaves the interface: we call this third contribution the transmission time, or transmission delay T_{tx}.

Finally, the *L* bits will travel serially along the A–B link: how long will it take before the first bit of the packet finally peeps out at interface B, followed by all other bits?

We call this last contribution the propagation delay T_p. It depends on two factors: the covered distance *d* and the speed *s* of the signal across the transmission medium. The dependency on distance is intuitive: if the point-to-point link between the two routers spans a few hundred meters, or if, on the contrary, it is an intercontinental link, several thousands of kilometers long, that makes quite a difference; it is straightforward to expect a more substantial propagation delay in the latter case. As for the medium dependence, the speed *s* at which signals propagate along the link varies for different media. Regardless of its actual value, T_p is given by the ratio

$$T_p = d/s. \qquad (1.11)$$

Note that T_p is merely a transit time, and as such, it has nothing to do with the speed at which bits are injected into interface A.

Summing up the four distinct contributions gives the overall delay, T_t, that the packet experiences in covering the considered hop between the two routers,

$$T_t = T_{proc} + T_q + T_{tx} + T_p. \qquad (1.12)$$

Fig. 1.16 pictorially represents the constituent terms of T_t: in this figure time *t* runs vertically, the shaded rectangles indicate the transmitted packet and their height is the transmission delay T_{tx}, or the packet time, as it is sometimes called.

Now enlarge the view, and draw the entire path covered by the packet moving from the source to its intended destination, as highlighted in Fig. 1.17. A LAN acting as the access network could be traversed, followed by various additional links, shared with

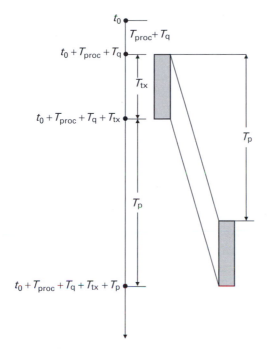

Fig. 1.16   A pictorial view of the different delays incurred by a packet on a single link

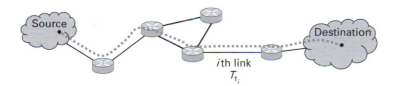

Fig. 1.17   The entire path covered by the packet

other packets. Even if at the end of each link we do not find a router, yet the conceptual classification of the delay contributions provided above holds.

Let us, therefore, assume that the packet crosses N hops before reaching its destination, and introduce an index i to tag the ith hop, so that the symbol T_{t_i} indicates the delay the packet experiences on it: then the total packet delay is

$$T_t = \sum_{i=1}^{N} T_{t_i} . \tag{1.13}$$

A related measure, often encountered in networks, is the round-trip time, commonly referenced by its acronym, RTT: it is simply defined as the time it takes for a packet to go from the source to the destination and then to return back to the source. In other words, the RTT is made of two one-way terms of the type detailed by Eq. (1.13).

1.7   A last miscellanea of concepts

1.7.1   Traffic sources

Unfortunately, data flows that traverse a network cannot be described in deterministic terms. If it were so, network design and planning would be quite simple: the only indication required to dimension the network "pipes" would be the transmission rate, in bit/s, of each data source.

Rather, network traffic sources are statistical in nature, and a wide range displays a bursty, intermittent nature: short time windows of intense activity are followed by prolonged intervals of no or reduced activity.

The burstiness factor B is the indicator that captures the most salient feature of a data source: it quantifies the fraction of time during which the source is actually emitting data.

The $B = 1$ value corresponds to constant bit rate (CBR) sources; in contrast, B values strictly lower than 1 identify variable bit rate (VBR) sources.

If reference is made to the simple case where a source in the active state emits data at a constant pace, note that $B = 1$ identifies the case where the source average transmission rate numerically coincides with the peak transmission rate.

A conventional voice coder is a typical example of a source producing CBR traffic. A client–server session generating traffic over the Internet is an instance of a VBR data source.

The picture of source traffic characterization is in reality much more complex: the few examples reported next should convince the reader of this.

Modern video coders, although CBR sources, can adaptively modify their output rate depending on the wealth of the uncompressed video information they are currently handling. Internet telephony does indeed imply the usage of a voice codec: yet, voice over IP (VoIP) coders exploit the presence of silence in between uttered sentences to produce digital voice samples only during the activity periods and, therefore, generate VBR traffic.

To conclude this rapid overview of traffic sources, we can state that, broadly speaking, CBR traffic is well served by circuit-switched technology and by data networks employing virtual circuits; on the contrary, packet-switched networks were introduced to handle VBR traffic: the concept of statistical multiplexing they are built on excellently meets the variability of this type of traffic.

1.7.2   Service taxonomy

We now shift our attention from the network infrastructure to the services that use it, and observe that there are alternative ways to categorize them: each draws on one peculiar feature of the application; in turn, this characteristic often has specific implications on the requirements that the underlying network is required to fulfil.

Talking about *configuration*, network services can be classified into:

- Point-to-point;
- Multipoint;
- Broadcast.

In a point-to-point service, two single endpoints – being actual users or processes – are involved and exchange information through the network. A file transfer application is an example of a service of this kind (peer-to-peer file sharing is excluded!); another common instance has to do with the fetching of web pages, downloaded from the server (the information repository) to the client (the information consumer).

In contrast, a multipoint service features the presence of more than two users, and requires the information that any user generates to be delivered to the whole group. Video conferencing is the most immediate example that comes to mind in this category. Definitely, a more challenging scenario than the previous category.

Last come broadcast services: they entail the presence of a single source and of a plurality of destinations. The broadcasting of television channels is the example we choose for this service class. Although dealing with a multiplicity of recipients, the implementation of a broadcast service is less demanding than the multipoint case, as the totality of the users has now to be reached.

Moreover, services can be distinguished as unidirectional or bidirectional: in unidirectional services, information flows in one direction only, from source to destinations; in bidirectional services, data are moved in both directions between the endpoints.

Further, bidirectional services can be symmetric or asymmetric: a telephone call, either supported via PSTN, or over the Internet, exhibits the first property; fetching a web page shows the second, as the underlying network application consumes much more bandwidth for the page download than for its request.

Definitely more significant, from a conceptual viewpoint, is the distinction of services into the three following categories:

- Delay-sensitive;
- Loss-sensitive;
- Bandwidth-sensitive.

Applications that fall within the delay-sensitive class do not tolerate the network to introduce an excessive delay in the communication process: if this unfortunately happens, the quality perceived by the user significantly deteriorates. Multimedia services are typical delay-sensitive applications. Video conferencing and IP telephony are among the most demanding services: if some of their information packets arrive too late, missing the playback deadline, they are useless and are equivalent to packet losses. For typical audio communications on a packet-switched network an overall delay T_p in the 0–150 ms range is acceptable; beyond the 350–400 ms limit, it is unbearable.

A delay-sensitive application can, on the contrary, stand a – limited – percentage of packet losses, typically of the order of 1%. This will reflect in a more modest quality (e.g., occasional glitches in audio and video signals), but the service will overall stay in place.

This is not so for loss-sensitive applications, which require every single information bit, from the first up to the last, to be correctly received. File transfer, electronic mail and Internet banking applications lie within this category.

Interestingly, some services within this category can endure substantial delivery delays.

As for bandwidth-sensitive services, they require a minimum value of *layer-2* bandwidth to be guaranteed in order to be successfully employed: the delivery of a high-definition video to a user connected to the Internet via a modest dial-up modem is clearly impossible; but even the transfer of a reasonable amount of data can turn into an extremely difficult task for this same user.

In other words, we can run into bandwidth-sensitive applications that either reside in the delay or the loss-sensitive realm.

1.7.3   Performance metrics

Having described the different types of service we are interested in, it is important to understand what performance metrics we adopt to assess the network's successful deployment.

The overall delay incurred by the information packets in traversing the network, in its constituent terms, detailed in Section 1.6, is the parameter for carefully monitoring delay-sensitive applications.

The packet-loss rate that the network introduces, for several different reasons (e.g., a temporary overload in one of its portions) directly relates to the performance of loss-sensitive applications. Such a rate is defined as the ratio of the number of packets that were not correctly received or went lost to the total number of packets that were sent over the network.

Moreover, networks can be employed in a more or less effective manner: throughput is the magic word to quantify the efficiency achieved, as its value measures how well we employ the available bandwidth.

Throughput can be defined differently in different contexts: for a single communication channel, a formal and accurate definition of throughput S is "the expected number of correctly received packet transmissions per packet transmission time," so that $S \leq 1$.

In common language, throughput is also meant to be the effective transfer rate an application experiences over a given layer-1 bandwidth: in this case *application throughput* is the more appropriate term.

No matter what definition we stick to, the indication that throughput provides is of great importance: a wireless LAN declaring a 54 Mbit/s layer-1 bandwidth typically achieves an application throughput that lies in the 20–25 Mbit/s range, and it is this value that ultimately tells us what "network speed" the service will experience.

Performance metrics appear and play a relevant role in the contractual terms that tie network users to their network service providers.

Among these terms there appear the so-called service level agreements (SLAs), documents that should identify the customer's need and, more specifically, define the services to be delivered, along with the performance goals set.

As an example, let us imagine a wide area network, fiber-based, interconnecting different sites, managed and supervised via a network operation center that continuously monitors devices and links.

Typical performance metrics that have to be tracked for each network device interface are the following: packet loss rate; available bandwidth; round-trip time from

Table 1.1. Service micro goals

Parameter	Goal
Datagram loss rate	< 0.1%
Guaranteed bandwidth	> 90% of the available bandwidth
RTT	< 20 ms
Jitter	< 10 ms

Table 1.2. Service macro goals

Parameter	Goal
Interface unavailability time	< 6 hours a year
Time to solve anomaly	< 6 hours in all cases
Average unavailability time per interface	< 2 hours

the interface of the examined network device to the operation center; jitter, i.e., RTT variation.

The goals to achieve regarding such metrics could be those described in Table 1.1, whereas the macro services guaranteed at each interface could obey the characteristics reported in Table 1.2: with reference to the latter table, a network interface is defined as unavailable when either the packet loss rate or the guaranteed bandwidth do not respect the corresponding limits.

1.7.4 Congestion and QoS

Strictly related to the delay and throughput metrics is the notion of congestion, an undesired network phenomenon that manifests itself when the traffic load becomes excessive.

When congestion occurs, throughput S rapidly decreases and delay increases, as Figs. 1.18 and 1.19 qualitatively indicate. Network resources are overwhelmed by packets: routers have buffers full of packets, and begin to discard them. Appropriate preventive actions are required to prevent congestion. If this is not possible, at the very worst congestion has to be confined.

Returning to the earlier example of the fiber-based WAN, different priority classes could be assigned to the outgoing traffic from any router interface: when the guaranteed bandwidth value is trespassed and congestion dangerously approaches, this would allow to drop low priority packets first.

In general, the introduction of priorities allows the network to carry delay-sensitive applications successfully even in the presence of high traffic levels, and more generally it guarantees different levels of quality of service (QoS) to heterogeneous network applications.

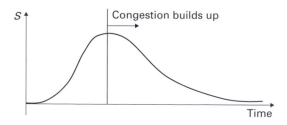

Fig. 1.18   Throughput as congestion appears

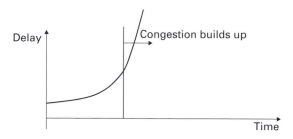

Fig. 1.19   Delay as congestion appears

1.8   A few bibliographical notes

This chapter owes a lot to the books that have set the path in the networking field: from the very beginning, *Telecommunication Networks* by Mischa Schwartz [1], *Data Networks* by Bertsekas and Gallager [2] and the lively *Computer Networks* by Tanenbaum [3], to the recent and innovative *Computer Networking* by Kurose and Ross [4].

The intent of the writers here is nothing but to introduce classical foundation concepts, functional to the treatments the next chapters will put forth, in a personal and hopefully agile style.

1.9   Practice: determining the RTT

Although we still remain in the foundations chapter, we can quantitatively consolidate a few of the concepts introduced earlier. We therefore mention two simple programs that are very popular among the Internet community: *ping* and *traceroute*.

Let us begin with ping: this command, originally introduced for Internet connectivity troubleshooting, allows one to verify whether a host belonging to an IP network is reachable. Its philosophy is rather simple: it sends a special request packet to the destination host, which in principle must accept it and respond with a special reply packet.

The request–reply packet exchange allows the round-trip time RTT between the host issuing the command and the "pinged" host to be estimated; in addition, ping records any packet loss, and prints a statistical summary when its execution comes to an end.

Ping is a very popular tool available on any major operating system, including Linux, Unix, MAC and Windows.

So, let us put it to work, invoking it from the command line of a Linux box toward another host, whose address is the "mysterious" 10.0.0.33 (the mystery will be disclosed in Chapter 2), which belongs to the same LAN. The -c option allows to specify the number of requests to be sent, five in our case.

```
[user@linuxbox ~]$ ping -c 5 10.0.0.33
PING 10.0.0.33 (10.0.0.33) 56(84) bytes of data.
64 bytes from 10.0.0.33: icmp_seq=1 ttl=64 time=0.121 ms
64 bytes from 10.0.0.33: icmp_seq=2 ttl=64 time=0.108 ms
64 bytes from 10.0.0.33: icmp_seq=3 ttl=64 time=0.096 ms
64 bytes from 10.0.0.33: icmp_seq=4 ttl=64 time=0.101 ms
64 bytes from 10.0.0.33: icmp_seq=5 ttl=64 time=0.088 ms

--- 10.0.0.33 ping statistics ---
5 packets transmitted, 5 received, 0% packet loss, time 3999ms
rtt min/avg/max/mdev = 0.088/0.102/0.121/0.017 ms
```

The command output shows that the destination replies to all five requests. Each displayed line reports, among other things, the RTT measured for the corresponding request. The last two lines report statistics about the number of packets sent, received and lost and the resulting minimum, maximum and average RTT values, as well as its estimated standard deviation. In this example, each request generates an RTT of the order of 100 μs, quite a small value justified by the considered set-up, where the link capacity is $C = 100$ Mbit/s and the physical distance between the two hosts is a few meters. Considering that the actual frame size transmitted on the wire is $L = 110$ bytes, the resulting transmission time is $T_{tx} = 8.8\ \mu$s, whereas the propagation time T_p is of the order of a few nanoseconds. We can, therefore, deduce that in this case the RTT is mainly due to packet processing and queuing delays, which are confirmed to be random, since we obtained five different RTT values even though packet size, bandwidth and distance were kept constant.

The role of the transmission time can be easily tested by increasing the packet size using the -s option to specify the payload length. If the latter is 1472, then the overall packet size becomes $L = 1526$ bytes (recall that a packet has both payload and control bits), so that $T_{tx} = 122\ \mu$s.

```
[user@linuxbox ~]$ ping -c 5 -s 1472 10.0.0.33
PING 10.0.0.33 (10.0.0.33) 1472(1500) bytes of data.
1480 bytes from 10.0.0.33: icmp_seq=1 ttl=64 time=0.351 ms
1480 bytes from 10.0.0.33: icmp_seq=2 ttl=64 time=0.349 ms
1480 bytes from 10.0.0.33: icmp_seq=3 ttl=64 time=0.347 ms
1480 bytes from 10.0.0.33: icmp_seq=4 ttl=64 time=0.339 ms
1480 bytes from 10.0.0.33: icmp_seq=5 ttl=64 time=0.350 ms

--- 10.0.0.33 ping statistics ---
5 packets transmitted, 5 received, 0% packet loss, time 4002ms
rtt min/avg/max/mdev = 0.339/0.347/0.351/0.012 ms
```

The measured RTT is now clearly affected by the transmission time, as more than two-thirds of the RTT are spent for the request and reply packet transmissions. The remaining variable delay is again caused by packet processing and queuing and is of the same order of magnitude as in the previous case.

Larger transmission times can be achieved by artificially reducing the link bit rate to $C = 10$ Mbit/s with the `ethtool` command. In this case the transmission time becomes ten times larger than in the previous case and it is definitely the major cause of the RTT delay.

```
[user@linuxbox ~]# ethtool -s eth0 speed 10
[user@linuxbox ~]# ping -c 5 -s 1472 10.0.0.33
PING 10.0.0.33 (10.0.0.33) 1472(1500) bytes of data.
1480 bytes from 10.0.0.33: icmp_seq=1 ttl=64 time=2.57 ms
1480 bytes from 10.0.0.33: icmp_seq=2 ttl=64 time=2.56 ms
1480 bytes from 10.0.0.33: icmp_seq=3 ttl=64 time=2.56 ms
1480 bytes from 10.0.0.33: icmp_seq=4 ttl=64 time=2.56 ms
1480 bytes from 10.0.0.33: icmp_seq=5 ttl=64 time=2.57 ms

--- 10.0.0.33 ping statistics ---
5 packets transmitted, 5 received, 0% packet loss, time 4001ms
rtt min/avg/max/mdev = 2.563/2.568/2.577/0.064 ms
```

The effect of the propagation delay becomes apparent when ping is directed to hosts geographically distant. In the following examples two ping sessions are run on a host in Italy toward servers located in California and Australia, respectively.

```
[user@linuxbox ~]# ping -c 5 -s 1472 www.berkeley.edu
PING www.w3.berkeley.edu (169.229.131.81) 1472(1500) bytes of data.
1480 bytes from 169.229.131.81: icmp_seq=1 ttl=43 time=197 ms
1480 bytes from 169.229.131.81: icmp_seq=2 ttl=43 time=197 ms
1480 bytes from 169.229.131.81: icmp_seq=3 ttl=43 time=197 ms
1480 bytes from 169.229.131.81: icmp_seq=4 ttl=43 time=198 ms
1480 bytes from 169.229.131.81: icmp_seq=5 ttl=43 time=197 ms

--- www.w3.berkeley.edu ping statistics ---
5 packets transmitted, 5 received, 0% packet loss, time 4001ms
rtt min/avg/max/mdev = 197.514/197.715/198.012/0.441 ms

[user@linuxbox ~]# ping -c 5 -s 1472 www.usyd.edu.au
PING solo.ucc.usyd.edu.au (129.78.64.24) 1472(1500) bytes of data.
1480 bytes from 129.78.64.24: icmp_seq=1 ttl=235 time=340 ms
1480 bytes from 129.78.64.24: icmp_seq=2 ttl=235 time=340 ms
1480 bytes from 129.78.64.24: icmp_seq=3 ttl=235 time=339 ms
1480 bytes from 129.78.64.24: icmp_seq=4 ttl=235 time=331 ms
1480 bytes from 129.78.64.24: icmp_seq=5 ttl=235 time=340 ms

--- solo.ucc.usyd.edu.au ping statistics ---
5 packets transmitted, 5 received, 0% packet loss, time 4001ms
rtt min/avg/max/mdev = 331.273/338.609/340.906/3.756 ms
```

Sometimes executing the ping command toward an active server does not provide the expected result, as in the following example.

```
[user@linuxbox ~]# ping -c 5 www.unimore.it
PING www.unimo.it (155.185.2.21) 56(84) bytes of data.

--- www.unimo.it ping statistics ---
5 packets transmitted, 0 received, 100% packet loss, time 3999ms
```

In this case all requests time out and no reply is ever received. We said earlier that a pinged host should accept the request and answer to it, but this behavior is not mandatory. Sometimes the reply is intentionally disabled, mainly for security reasons. Furthermore, some ISPs filter out request messages too to avoid attacks and confine traffic on their networks.

The RTT increases with the physical distance, although all its components play their role, since processing, queueing and transmission are performed on each node along the path to the destination.

An interesting hop-by-hop breakdown analysis of the RTT, especially when the two hosts are quite distant, is provided by the traceroute command. This tool allows one to determine the route followed by IP packets across an IP network and records the contribution of each hop to the overall RTT; as with ping, it is available on practically all operating systems.

The following example lists the hops along the path from a host in Bologna, Italy, to one in Sydney, Australia, along with the corresponding timing information, namely, three measured delays.

From the measured values it is evident that the intercontinental link is located between nodes 10 (62.40.112.22, so-7-2-0.rt1.fra.de.geant2.net) and 11 (202.158.204.249, ge-3-3-0.bb1.a.fra.aarnet.net.au), connecting the European research and education network (geant2) to the homologous Australian network (aarnet).

```
[user@linuxbox ~]# traceroute -n -m 23 www.usyd.edu.au
traceroute to www.usyd.edu.au (129.78.64.24), 23 hops max, 40 byte
 1   192.168.8.254   0.347 ms    0.344 ms    1.513 ms
 2   137.204.191.254   1.773 ms    0.455 ms    0.550 ms
 3   137.204.20.125   0.884 ms    0.837 ms    1.073 ms
 4   137.204.47.62   2.205 ms    2.350 ms    2.071 ms
 5   137.204.2.17   2.712 ms    2.603 ms    2.699 ms
 6   193.206.128.125   3.209 ms    2.727 ms    2.701 ms
 7   193.206.134.21   6.332 ms    5.971 ms    6.245 ms
 8   62.40.124.129   6.074 ms    6.206 ms    6.345 ms
 9   62.40.112.209   13.439 ms    13.452 ms    13.239 ms
10   62.40.112.22   21.965 ms    21.611 ms    21.348 ms
11   202.158.204.249   206.959 ms    207.084 ms    206.726 ms
12   202.158.194.145   206.989 ms    206.859 ms    209.330 ms
13   202.158.194.129   295.273 ms    290.924 ms    255.604 ms
14   202.158.194.6   282.799 ms    329.899 ms    291.295 ms
15   202.158.194.18   335.398 ms    342.511 ms    327.393 ms
16   202.158.194.178   293.783 ms    295.157 ms    332.879 ms
17   202.158.194.30   313.871 ms    305.254 ms    304.522 ms
```

```
18   202.158.194.201   303.387 ms    311.038 ms    317.405 ms
19   202.158.194.42    314.747 ms    338.956 ms    317.223 ms
20   202.158.202.194   363.602 ms    361.542 ms    357.560 ms
21   202.158.202.202   337.367 ms    339.412 ms    336.243 ms
22   * * *
23   129.78.128.29 (N!)   330.357 ms * *
```

1.10   Exercises

1.1   Q – Let us face the important issue of how to identify different virtual circuits within a network. Try to spot the right solution.

A – Intuition might erroneously lead us to think that the virtual circuit numbering obeys a global scheme, common to the entire network, and that once the circuit has been set up and identified via a number, it is this number to be placed in the control field of every packet to allow correct routing decisions at each node. Not quite so: having a unique (numerical) identifier for each circuit would require all routers to know and agree on it; in turn this would result in the exchange and processing of a significant number of messages among routers, quite an awkward solution. What happens is that the virtual circuit is identified by a different number on each of the links it crosses: this is possible as every router replaces the virtual circuit number in the control field of the incoming packet with a new number, before forwarding the packet along the correct outgoing link. To perform this task, the router employs a translation table, which is built ad-hoc when the virtual circuit is originally set up. To understand how this numbering actually works, Fig. 1.20 shows a simple example, the continuous line representing the route of the examined virtual circuit; assume that it is identified by the following sequence

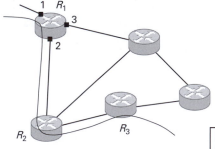

Input interface	Input VC number	Output interface	Output VC number
1	24	2	19
2	43	1	16
3	8	2	15
1	86	2	10

Fig. 1.20   A virtual circuit example and the translation table of router R_1

of numbers: 24, 19, 23. *Let us now take a look at a portion of the translation table of router R_1 of the example, also reported in Fig. 1.20. To explain it, note that router R_1 has three interfaces, numbered 1, 2 and 3. The first highlighted entry refers to the examined virtual circuit, and indicates that on the interface where this virtual circuit enters the router, interface 1, 24 is the number identifying it; this number is 19 on the outgoing interface, interface 2. These will be the numbers used to handle the packets belonging to the virtual circuit properly: in other words, all packets entering router R_1 on interface 1 with virtual circuit identifier 24 in their control field will see their identifier changed to 19; then they will be diverted to interface 2. How many entries does a translation table possess? As many as the virtual circuits that cross the router. These vary in time, so that the translation table has to be dynamically updated, inserting new entries and removing the outdated ones. Quite a laborious task, indeed.*

1.2 **Q** – Let us consider M nodes that have to be interconnected via a flat topology network. How many links do the different solutions require? What can we say about the average distance, given in terms of traversed links, between any two stations?

A – *The bus requires the adoption of one common link (plus M interfaces to the link, one for each node) and the distance between any pair of nodes is 1. The star needs M links (plus the star hub placed in its center); the distance between nodes is constant and equal to 2. The ring requires M links to be constructed, the distance lies between 1 and $\lfloor M/2 \rfloor$. The tree is built via $M - 1$ separate links. The fully connected mesh uses $M(M - 1)/2$ links and the distance between nodes is constant and equal to 1; the partly connected solution employs X links, with $M < X < M(M - 1)/2$: the average distance between nodes depends on the actual positions of the available branches between nodes.*

1.3 **Q** – The simplest model of a bursty traffic source assumes that ON and OFF periods of constant duration alternate along the time axis: during the ON periods the source is emitting bits at a given bit rate B_r, during the OFF periods it is silent. Draw its output rate as a function of time and determine the corresponding burstiness factor .

A – *The behavior of the source is graphically described in Fig. 1.21. Given that T_{ON} and T_{OFF} represent the duration of the ON and OFF periods, respectively, the source burstiness factor is given by $B = \frac{T_{ON}}{T_{ON} + T_{OFF}}$.*

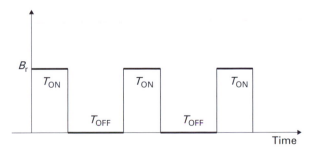

Fig. 1.21   Emitting data rate as a function of time for the simplest ON–OFF source

1.4   Q – Let us consider an ON–OFF source, modeled via the state diagram of Fig. 1.22, where μ and λ represent the rate at which the source leaves the ON and the OFF state, respectively. Moreover, let us assume that the time the source spends in each state is exponentially distributed and that in the ON periods its bit rate is B_{r}.

(a) Find the probability P_{ON} that the source is in the ON state.
(b) Determine the source burstiness factor.

A – Once the source reaches the equilibrium working point, given that P_{OFF} represents the probability of the source not emitting data, we have that $\lambda \cdot P_{\mathrm{OFF}} = \mu \cdot P_{\mathrm{ON}}$ holds: this relation quantifies the equilibrium law affirming that, in each state, the flow of probability entering the state must equal the probability flow out of the same state. This relation and the normalization condition $P_{\mathrm{ON}} + P_{\mathrm{OFF}} = 1$ form a system of two linear equations in the unknowns P_{ON} and P_{OFF}: the corresponding solution is $P_{\mathrm{ON}} = \frac{\lambda}{\lambda+\mu}$ and $P_{\mathrm{OFF}} = \frac{\mu}{\lambda+\mu}$. The source burstiness factor is $B = P_{\mathrm{ON}}$, as P_{ON} can also be interpreted as the percentage of time the source spends emitting bits.

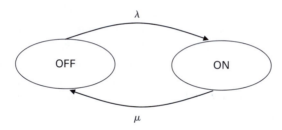

Fig. 1.22   State diagram of the ON–OFF source with exponentially distributed times in both states

2 Architectures and protocols

This chapter takes the reader into the modern world of computer networks: it first illustrates the main standardization bodies for telecommunications and data networking, then it exposes the general principles that underlie communications in a packet-switched network and describes the features of the main protocols ruling the proper functioning of the Internet.

An adequate, thorough treatment is devoted to the Internet protocol (IP), as its functionalities are the basis on which the entire Internet communication model relies: the original classful and the current classless addressing schemes are presented, followed by the subnetting and supernetting procedures IP allows. Direct and indirect datagram delivery is also introduced. Every topic is complemented by some explanatory examples, whose aim is immediately to translate into useful, working cases, the theoretical concepts.

The chapter is closed by two sections that present the salient attributes of the transmission control protocol (TCP) and of the user datagram protocol (UDP).

2.1   Who's who in the telecommunication world

It is appropriate to place in this opening section a brief, yet meaningful overview of the most representative bodies within the telecommunication and networking arena, and of their role. As the topics covered in the next chapters will quantitatively testify, the networking field heavily relies upon standards: they are needed to guarantee for systems and services not only quality, safety, reliability and efficiency, but also interoperability, a magic word when the ultimate objective is to "connect."

The International Telecommunication Union (ITU) [5] plays a leading role in standardization. The ITU is the leading United Nations agency for information and communication technologies. Based in Geneva, Switzerland, ITU membership includes 191 member states and more than 700 sector members and associates. It might be interesting to cite the most recent ITU activities related to the telecommunication and information technology fields at the time of this writing: they deal with the exciting field of IP-TV, and correspondingly consider multimedia application platforms, content formats and end systems. The ITU executive branches are called focus groups, their main task being to deliver documents that ITU will assimilate into its work program: the final intention is to generate standards, called ITU-T recommendations.

In Europe, the European Telecommunications Standards Institute, ETSI [6], further produces globally applicable information and communications technology (ICT) standards. The ETSI is a not-for-profit organization with almost 700 ETSI member organizations drawn from 60 countries worldwide, and it is recognized by the European Commission as a European standards organization. Some examples of ETSI popular standards are to be found in the field of digital terrestrial television, and hide behind the acronym DVB, meaning digital video broadcasting; also the transmission system for Hand held terminals (DVB-H) is covered by proper ETSI standards. ETSI had a prominent role in the 3rd Generation Partnership Project (3GPP), a collaboration agreement that led to the standards for the most popular existing mobile radio networks (GPRS and UMTS).

The International Organization for Standardization [7], ISO for short, derived its name from the Greek isos, meaning "equal," and produces standards too. ISO is a nongovernmental organization, bridging public and private sectors. Many of its members are part of the governmental structure of their countries; others have been set up by national partnerships of industry associations. ISO solutions should, therefore, meet both the requirements of business and the broader needs of society. ISO standards cover telephone networks, PSTNs and powerline telecommunications, just to cite a few. In the next section a salient, although aged, ISO standard will be introduced, still useful to logically frame the distinct issues to face when transmitting data over a packet-switched network.

Another historical developer of standards in both traditional and emerging fields is the Institute of Electrical and Electronics Engineers (IEEE) [8]. Its telecommunication standards mainly lie within the wired and wireless networking domain. There is no doubt that IEEE documents constitute the reference when describing how wired and wireless LANs operate. Also, the emergence of new technologies in the personal and metropolitan networking areas are promptly captured and reflected by the standards this body promulgates.

The IEC, the International Electrotechnical Commission [9], is the organization that prepares and publishes international standards for all electrical, electronic and relating technologies. Some of its standards embrace the telecommunication field, when questions relating to electrotechnology have to be answered.

Left as the last citation (last comes the important guest: isn't this the order followed at banquets, plenaries and parties?), is the Internet Engineering Task Force (IETF) [10]. Undoubtedly, when talking about networks, the lion's share is for the Internet: IETF is the community whose mission is

... to produce high quality, relevant technical and engineering documents that influence the way people design, use, and manage the Internet in such a way as to make the Internet work better. These documents include ... standards, best current practices, and informational documents of various kinds [11].

The actual technical IETF work is performed in its working groups, logically organized by topic (e.g., routing, transport, security). The Internet standards are then published in openly accessible documents, requests for comments (RFC), frequently cited in all subsequent pages. Every RFC is identified by a numeric value: the higher the number, the more recent the document.

More standards bodies do exist, and play a role when networking is considered: the previous listing was intentionally confined to international organizations, while nation-wide institutions were omitted. The reason is that the latter have among their institutional roles that of strengthening the marketplace position of the country they represent in the global economy. On the contrary, in this book the main emphasis is on the technical role that internationally recognized standards play, and on their importance in our daily lives, when we face communication needs. As a matter of fact, we trust a local area network to work as expected, a wireless interface card to be recognized and smoothly operate with any wireless access point, an email message to be read by an application either under Linux or under Windows, in Italy as in Korea.

2.2 OSI Model: the seven-layer approach

As promised, this section hosts the citation to a venerable, yet popular ISO standard. It is the ISO Model for Open System Interconnection (OSI), currently classified as "with-drawn," originally documented in ISO 7498 [12] and its addenda. The next paragraphs aim at depicting what its main contributions were, letting the reader understand what has survived out of it.

In essence, the ISO OSI Model can be interpreted as an excellent framework to under-stand logically all the issues to tackle and solve when dealing with data communication through a packet-switched network. Classifying them in an abstract manner, the OSI Model provides the novice with a picture of the information flow, clearly stating the fundamental concepts embedded in any communication architecture and into the cor-responding implementation. As Fig. 2.1 indicates, this model is organized into seven

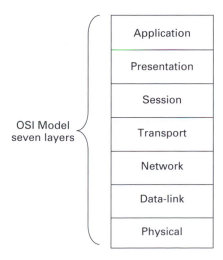

Fig. 2.1   The OSI reference model

abstraction layers: from the bottom to the top of the architecture, they are termed physical, data-link, network, transport, session, presentation and application layers (equivalently, they are identified as layer 1 through layer 7).

A concise read-through of the main functionalities of each layer leads to the description proposed hereafter.

For the physical layer, it can be roughly said that its goal is to transmit bit streams onto the underlying communication channels. The functions it performs are: activation and deactivation of physical connections, transmission and multiplexing of data units over the transmission medium. The layer-1 features of a LAN will detail, for example, the medium employed, the signal levels, the supported transmission rates, the modulation and coding techniques – if any – adopted.

The data-link layer is definitely more articulated, as a data-link connection can be built on one or several physical connections. From the experience gained in the years that followed the OSI Model definition, it is appropriate to affirm that this layer currently faces themes in two logically distinct areas.

The first area embraces communications occurring on a point-to-point link, when the two devices that wish to communicate sit at the extremes of the link. Here the goal of the layer depends on whether the underlying transmission medium is wired or wireless. In wired networks, and specifically in WANs, point-to-point links do not require heavy functionalities at this layer: transmission media are reliable, and the rare errors that might be introduced at the physical layer are detected at the most, but not corrected. In this case data-link layer solutions are lightweight: their main tasks are addressing and data encapsulation. On the contrary, when a wireless link is examined, the goal of the layer is to convert the transmission medium, employed along the physical layer specifications, into a communication channel that is as close as possible to an error-free channel for the network layer above. At present, this target is of paramount importance over the wireless access link, be it in wireless LANs or MANs, but also in mobile radio networks: given the unreliability of the radio channel, the data-link layer has to mask its imperfections, typically through the adoption of appropriate retransmission policies.

The second area of interest for the data-link layer is encountered when considering more stations wishing to communicate via a shared channel. This topic fits one specific sublayer of the data-link layer, named medium access control (MAC), originally not foreseen by the OSI Model. Here the problem to solve is how to coordinate the stations, so as to guarantee all of them a fair access to the communication channel, and possibly few, if not any, conflicts when accessing the common channel. We will pick up again on this subject in the next chapter.

The last remark on the data-link layer deals with terminology: a layer-2 data unit is commonly named "frame," as opposed to "packet," which identifies a layer-3 data unit; we will diligently adhere to this convention throughout the text.

For the network layer, the keywords that summarize its main functionalities are: routing and relaying. This layer is concerned with moving the information end-to-end, from source to receiver. When datagram and virtual circuit networks were introduced in Chapter 1, reference was implicitly made to two alternative features that the network layer of a packet-switched network can exhibit: indeed, packets belonging to the same information stream can be routed in a datagram manner, or can follow the same virtual pipe.

The transport layer was originally intended to be the core of the OSI model. Its role? To warrant a transparent transfer of data from sender to receiver; if needed, the transport service it offers can be *reliable*. A duplicate task, with respect to the network and the underlying layers? Not quite, as reliability is important for some applications. Just think of a file transfer, and its requirement to have all bits correctly delivered through the network. In this case the transport layer has to act as a filter, hiding all the errors that the layers below could have missed or were not able to cope with.

In the light of the last statements, it is possible to interpret the four lower layers of the OSI Model as a unique transport service *provider*, and the remaining three upper layers as the *user* of such transport service [1].

Interestingly, note that the services the transport layer provides are implemented at end users only, whereas the physical, data-link and network layers are to be found in both end users and network devices. It is at the transport layer that the user – or we should better say the operating system of its machine – can perform an end-to-end control over the communication process.

As for the session and presentation layers, they guarantee several add-on services. If needed, the session layer can handle the establishment and the release of the session connection, it can distinguish between normal and expedited data transfer, it can resynchronize the session. In turn, the presentation layer is concerned with the problem of the common representation of information, i.e., with its syntax. It is, therefore, responsible for selecting mutually acceptable transfer syntaxes.

Finally, the application layer: it provides the sole means for the application process to access the network environment. All network applications sit within this layer, from the ones that offer a file transfer service, to those that allow one to send and receive mail messages.

How the information is moved through the network when the OSI Model is adopted is summarized in Fig. 2.2, which refers to the case of two end users – or open systems, as the OSI Model terms them – that are directly connected: for the sake of simplicity, we can imagine them belonging to the same network, say a LAN. The sender passes the information to its application layer, either in the form of a structured stream, or already divided into small chunks of bits (one of these chunks appears at the top of the figure). The application layer takes care of each single piece of information, adding a header and passing the whole new unit to the presentation layer below; the presentation layer interprets this block as a whole, not investigating what its content is, adds its own header and handles the resultant data unit to the session layer. This process continues, layer after layer, in a manner that recalls the operation of placing a letter within an envelope and the resulting dispatch in a further envelope, and so on. When the data-link layer is reached, typically both a header and a tail are added. What do all these headers and tail contain? Arcane as it might sound at the moment, they contain all the information needed to support the functionalities that the different layers provide. At the receiving end, a dual process is performed: headers are gradually stripped away – and interpreted – as the packet successfully goes up the stack. The key concept lying behind this mechanism

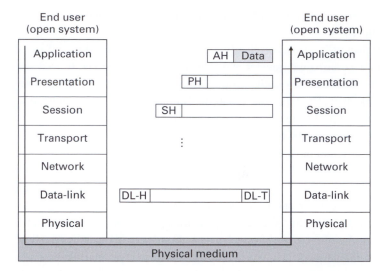

Fig. 2.2   The encapsulation and communication process across the seven layers

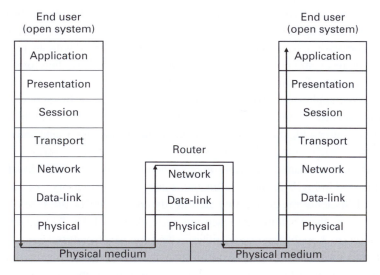

Fig. 2.3   Communication via a router

is that of *horizontal* communication: although the information flow is vertical, from the top to the bottom of the stack at the sender and from the bottom to the top at the receiver, the peer layers talk to each other. Explicitly, when the process at the network layer of the sending machine passes its data unit to the data-link layer below, it actually sends it to the network layer process on the receiving machine, that is, the ultimate beneficiary of the message.

A slightly more articulated example is reported in Fig. 2.3, where the case of two end users belonging to two different networks, say LANs, connected via a router, is

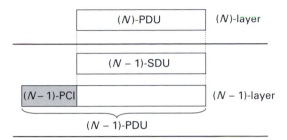

Fig. 2.4   Mapping between data units in adjacent layers

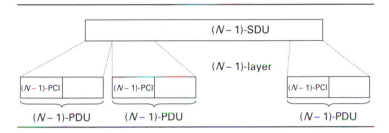

Fig. 2.5   Segmentation of one SDU into more PDUs

examined. This figure shows that the functionalities of the three lower layers are indeed to be implemented in routers too.

Conventionally, layer-$N - 1$ provides a *service* to layer-N, through its $(N - 1)$-layer *entities*, and makes the service accessible at the $(N - 1)$th layer *service access points*, $(N - 1)$-SAPs. Also, as Fig. 2.4 evidences, the block of information that the Nth layer works with is called the (N)-*protocol data unit*, (N)-PDU for short; once the (N)-PDU crosses the boundary between layer-N and layer-$N - 1$, it becomes an $(N - 1)$-SDU, where SDU stands for *service data unit*. The addition of proper control information to the latter, termed protocol control information (PCI), in the form of either a header, or a header and a tail, leads to the $(N - 1)$-PDU. The mapping between (N)-PDU and $(N - 1)$-SDU might or might not be one-to-one: segmentation can occur, as exemplified in Fig. 2.5, that indicates the mapping of one SDU into more PDUs. In this circumstance, all fragments of the original SDU have to be properly identified in the $(N - 1)$th PCIs, to allow the receiving end to reassemble the original SDU. This is what currently happens when PDUs transit from the network layer to the data-link layer of wireless networks. Conversely, the case of several $(N - 1)$-SDUs forming a single $(N - 1)$-PDU can also occur.

Moreover, four distinct service primitives rule the dialogue between adjacent levels: these are the *request*, *indication*, *response* and *confirm* OSI primitives. An example of how they are employed in a specific context will be illustrated at the end of this section.

The aged OSI Model has left us with one last, precious contribution: the introduction of three distinct concepts, namely: service, protocol and interface. What does service mean? As evidenced before, each layer provides one or more services to the layer below.

The description of a service, therefore, details *what* the layer does. The routing of the packets through the network, when the network layer is considered, for example.

How is the service actually performed? Through what specific modalities? The description of the service protocol answers this question. It is possible to state that a protocol defines the format of the messages that peer entities exchange, their order, and also defines the actions that follow the transmission and the reception of a data unit in a layer.

Last, the notion of interface: it provides the information about *where* service is made available, together with the definition of the required input and the expected outputs.

Now that these concepts have been introduced and clarified, the notion of *connection-oriented* and *connectionless* services can be put forth. A connection-oriented service exhibits three distinct logical phases: regardless of the layer involved, these are the connection-establishment phase, the data-transfer phase and the connection-termination phase. A connectionless service, on the contrary, displays the bare data-transfer phase. In current networks, the data-link, network and transport layers offer either connection-oriented or connectionless services. Some practical examples: the network layer provides a connectionless service in a datagram network; it offers a connection-oriented service in a virtual circuit network. A few additional examples are anticipated here, which will find a sound justification in the next chapters: the MAC sublayer in wired LANs provides a connectionless service; the MAC sublayer of the recent IEEE standard for wireless MANs is, on the contrary, connection-oriented. The transport layer of the Internet offers a connection-oriented service through the functionalities of its popular transmission control protocol (TCP), and a connectionless, lightweight service through its alternative protocol, the user datagram protocol (UDP).

Interestingly enough, several modern IEEE standards have inherited OSI terminology, and professionals often encounter the PDU, SDU, SAP terms, as well as the OSI service primitives, when striving to understand the packet structure and the dialogue between adjacent layers adopted in current networks. An example of this is reported in Figs. 2.6 and 2.7, both referring to the IEEE 802.16 standard for Wireless MANs.

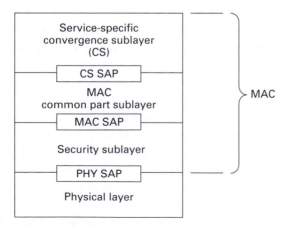

Fig. 2.6   Logical structure of the lower layers of an 802.16 Wireless MAN

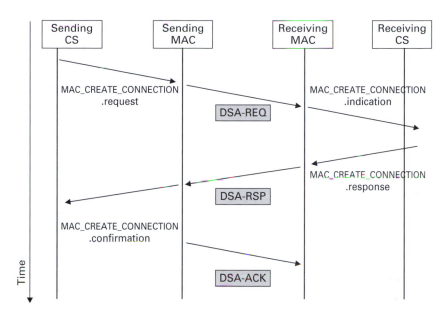

Fig. 2.7 The process of opening a MAC connection in an IEEE 802.16 Wireless MAN

The first figure details how the MAC sublayer is organized: as expected, life in reality is more complex than expected, and the IEEE 802.16 MAC encompasses a security sublayer, a MAC common-part sublayer and a service-specific convergence sublayer (CS). The latter figure illustrates the dialogue, ruled by the four OSI service primitives, between the CS and the MAC common part sublayer to open a MAC connection: also shown are the three frames, DSA-REQ, DSA-RSP and DSA-ACK (request, response and acknowledgment, respectively), successfully exchanged among the entities in the two sublayers.

2.3   TCP/IP protocol suite

Although the OSI Model introduced fundamental networking concepts and its architecture is still valuable today, OSI protocols never really took off. They were initially plagued by complexity and inefficiency, and could not rival the concomitant diffusion of the TCP/IP protocol suite, so called from the names of its two most popular protocols. It is inaccurate to use the word "communication model" with reference to this suite, which rules the communication over the Internet and which perfectly reflects the IETF approach: efficient, running code. Nevertheless, there is a model implicitly underlying it: it displays four layers, as Fig. 2.8 indicates, which, from bottom to top, are conventionally termed host-to-network . . . interface (isn't it strange to label a layer with the word "interface?"), Internet layer, transport layer and application layer. While the latter three resemble in scope the analogous OSI layers, the host-to-network interface, in quite a rough manner, is the place where all the "technicalities" that guarantee the connection of

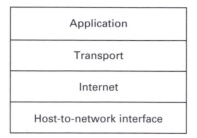

Fig. 2.8   The TCP/IP "model"

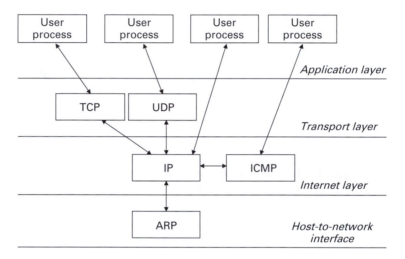

Fig. 2.9   TCP/IP protocols

the host – be it a PC, a workstation, or a generic device – to the Internet should be tackled and solved; its counterparts are the physical and data-link layers of the OSI Model. The word "technicalities" is downsizing: there are plenty of challenging engineering issues within the lower part of any communication architecture; some of the most innovative, disruptive services and systems of the last years were made possible thanks to great technical achievements reached in these lower layers: from ADSL to UMTS, from optical networks to high-speed routers. Nevertheless, we will stick to the original terminology, and unwillingly retain the awkward term "interface."

Back to the TCP/IP protocol suite; its description translates into commenting the main features of its most important protocols: the ones referenced in this chapter are shown in Fig. 2.9, which, at a glance, locates them in each layer and indicates their mutual relationships: the address-resolution protocol (ARP) within the host-to-network-interface, the Internet protocol (IP) and the Internet control message protocol (ICMP) at the Internet layer, the transmission control protocol (TCP) and the user datagram protocol (UDP) at the transport layer.

Within the Internet layer, IP is by far the most popular actor of the suite. At the source, its first responsibility is to encapsulate the payload it receives – typically from

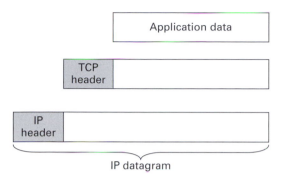

Fig. 2.10   Forming an IP datagram

the transport layer above – in a packet, the familiar *datagram* we first encountered in Chapter 1. The mechanism is the same as in the OSI model: Fig. 2.10 highlights the intermediate steps that lead to the formation of a datagram, when data are passed to the IP module by TCP. The motto to remember is: "everything that is transmitted over the Internet is carried in IP datagrams."

Once a datagram has been formed, the IP's crucial task is to route it to its final destination. Each device connected to the Internet is – in principle – identified by a unique IP address, used by the IP modules that are in charge of routing of the datagram: in a connectionless manner, ruled by the IP hop-by-hop philosophy, the datagram crosses the network from source to destination.

The IP is implemented both within hosts and routers – called *gateways* in the original Internet jargon – and offers a best-effort service, meaning that it does its best to deliver the datagram to its final destination; yet it offers no guarantees that this will happen. If something goes wrong, the IP only tries to inform the source that the datagram did not make it to the destination, via a proper control message of the Internet control message protocol (ICMP)[13]. No recovery procedures are triggered by this protocol; no retransmission policies are started.

At the transport layer, two alternative protocols are available: TCP and UDP, which display complementary features. What this actually means is that whenever an application requires an accurate delivery, TCP has to be employed, as this protocol handles the retransmission of those bytes that were not correctly acknowledged by the TCP receiving module. File transfers and mail services, to name two popular network applications, rely upon TCP. Moreover, TCP provides a connection-oriented service and successfully enforces congestion control.

On the contrary, UDP offers a transport service alternative to TCP, with no guarantees of correct delivery; its service is connectionless, and UDP enforces no congestion control. This protocol is employed by those applications that privilege fast to accurate delivery, such as real-time voice and video applications.

Above these two protocols, all the application realm is to be found: the original telnet and the file transfer protocol (FTP) reside here, now paired by a plethora of new Internet services, spanning from peer-to-peer for file sharing and video streaming to voice over the Internet (VoIP) applications. Unlike in the OSI Model, in the TCP/IP protocol suite the application layer encompasses everything left out of the delivery process accomplished by the layers below: no trace of session and presentation layers is found in the Internet.

The remaining part of this chapter will explore the salient features of the IP, TCP and UDP protocols; an explanation of the mechanisms ARP employs will, on the contrary, be postponed to the next chapter.

2.4   IP: Internet protocol

As RFC 791 indicates [14], IP implements two basic functions: addressing and fragmentation. Addresses in each IP datagram are required to understand who is the source and who is the final recipient of the transmitted data: it is the destination address that is the most important ingredient needed to route the datagram across the network, and it is this address that is used to drive the selection of the path in the IP modules of the routers generally traversed by the datagram. These modules share common rules for interpreting address fields and implement procedures to take proper routing decisions.

As for fragmentation of a datagram, this is necessary when the datagram, to reach its destination, has to traverse a network where the packet size limit is smaller than in the originating network. Proper fields within the IP datagram allow for the handling of the fragmentation and defragmentation procedures.

Overall, IP provides a connectionless service, as each datagram is routed independently of other datagrams belonging to the same information flow. There is no ambition to provide a reliable service: hence, IP employs no acknowledgments, either end-to-end or hop-by-hop, and performs no retransmissions. Moreover, it enforces no flow control.

The header of a generic IP datagram is reported in Fig. 2.11, where each row represents a word, i.e., 32 bits, of the datagram. The order of transmission of the header and of the data following it is the normal order in which they are read in English: from left to right, one byte after the other; from top to bottom, row after row.

The first four-bit field provides the version of the IP header format: the description that follows refers to version 4. Then comes the header length field (HLF): its value gives the length of the IP header, in words of 32 bits: the minimum value is five, corresponding to 20 bytes. This is also the default IP header length, exhibited by the majority of the datagrams circulating over the Internet.

The next field is called "type of service" (TOS): it is eight bits long, and specifies what treatment the datagram should undergo as it traverses the network. After the original promulgation of RFC 791 [14], RFC 1122 [15] and RFC 1349 [16] have better detailed the usage of the TOS field. Following their indications, bits 0–2 of this field constitute the precedence subfield, i.e., denote the importance or priority of the datagram. Commercial routers currently implement various kinds of queueing services, including priority queueing, based on policies encoded in their filters: the IP precedence field is among the options such filters can examine. To complete the story, bits 3–6 provide an indication of the trade-off between the delay, throughput, reliability and cost that the

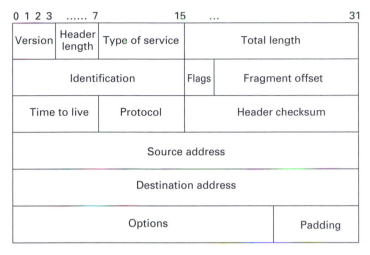

Fig. 2.11   IP header

Fig. 2.12   IP header: a detail of the flags field

datagram can bear: to make things clear, the 1000 sequence for bits 3–6 means "minimize delay," and analogous combinations tag the alternative goals: 0100 stands for "maximize throughput," 0010 for "maximize reliability" and 0001 for "minimize cost." Bit seven is left unused for future specifications. The TOS default value is all zero bits. Subsequently, RFC 2474 [17] has been the first RFC to elaborate further on the crucial topic of service priority for different datagrams: it proposes a completely new layout of the TOS octet, in view of guaranteeing different levels of quality of service (QoS) to different streams of IP datagrams: the interested reader is encouraged to dig for further details in it.

The total length field gives the datagram length, measured in bytes, including header and data. Its size, two bytes, indicates that the maximum, theoretical length is 65535 bytes. No network can actually handle such huge datagrams: typically, far smaller units circulate over the Internet, as Chapter 3 will show.

The identification field holds a value assigned to the datagram by the sender: it will aid in assembling the fragments of a datagram, if this action is needed. The next field is called flags and is three bits wide: it is organized as detailed in Fig. 2.12. Bit 0 is reserved and must be zero; bit 1 is the DF bit: set to 0 means "may fragment," set to 1 means "don't fragment"; bit 2 is the MF bit: set to 0 indicates "last fragment", set to 1 indicates "more fragments." The 13-bits fragment offset field completes the set of tools needed when fragmenting and defragmenting: it indicates where in the datagram this fragment belongs; the offset is measured in units of 64 bits, the first fragment having offset 0.

Time to live (TTL): this field reports the maximum lifetime allowed for the datagram. The rules are simple: the TTL value is decremented by one each time the datagram crosses a router; moreover, the datagram is discarded when its TTL reaches the zero. The purpose of this field is to avoid the possibility that undeliverable datagrams indefinitely keep circulating over the Internet.

The protocol field: it indicates the protocol used by the sender to pass its data to the IP module for encapsulation. Such information turns out to be useful at the receiver, when the payload of the IP datagram has to be handed to the appropriate protocol module, e.g., TCP. The values that identify different protocols were originally specified in RFC 820 [18].

Now the header checksum field: it "protects" the datagram header only, enabling the receiving device to verify whether the header has been correctly received or has been altered. It is easily computed: first of all, 16-bit words of the header are summed, then the ones' complement of the result is evaluated to determine the checksum via a proper algorithm. As some header fields – e.g., the time to live – are modified along the route, the checksum needs to be recomputed and dynamically updated by any device that changed any field in the IP header.

The source and destination addresses, four bytes long each, are loaded in the next two fields: the addressing scheme that the IP protocol employs deserves an adequate explanation, as the next subsection will present.

The options and padding fields close the IP datagram header. The former is of variable length, and encompasses some – optional – tools to monitor the proper functioning of several network functionalities, namely: the loose source routing option forces the datagram to cross a given list of IP devices to reach its destination; the strict source routing forces the datagram to cross exclusively the given list of IP devices; the record route option is used to trace the route the datagram takes; the timestamp option requires each traversed router to record (append) its IP address and time to the datagram.

The last is an all-zero field, called a padding field, and is used to guarantee that the header ends on a 32-bit boundary, adding as many zeros as needed.

2.4.1   Public and private IP addressing

One of the main tasks of any network-layer protocol is to define a global addressing scheme, allowing each network element to be uniquely identified. We just outlined that version 4 of the Internet protocol, IPv4 in what follows, adheres to this statement, putting forth an address space based on fixed-sized addresses of four bytes. Internet protocol addresses are traditionally expressed in *dotted decimal* notation: each of the four octets can be represented by a decimal number in the 0–255 range. For instance, the address

$$10001001.11001100.11010100.00000001$$

is given as 137.204.212.1.

Fig. 2.13   Structure of an IPv4 address

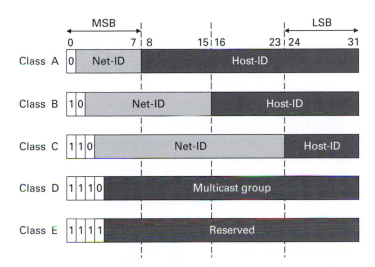

Fig. 2.14   IPv4 classful addressing scheme

In theory, the choice of using a 32-bit field would make more than four billion IP addresses available. However, to avoid complete anarchy in picking up addresses from such a big, flat space, with the consequent, harsh problem of keeping track of the assignments, IPv4 developers wisely decided that assignments should be hierarchically performed, grouping contiguous addresses into *IP networks*.

The structure of the IP address reflects this approach: it exhibits a network part, identifying the IP network the address belongs to (*net-ID*), and a host part, identifying the host within the network (*host-ID*), as shown in Fig. 2.13. The higher the number of bits dedicated to the host-ID, the higher the number of hosts that can be included in a network.

IP networks were originally divided in three classes, based on their size:

Class A   networks with a large number of hosts;
Class B   networks with an intermediate number of hosts;
Class C   networks with a small number of hosts.

More precisely, class A reserves three bytes to the host-ID, class B two bytes and class C a single byte. Some time later, RFC 1112 [19] defined *class D*, including addresses identifying host groups for multicast transmission, and *class E*, comprising addresses reserved for future use. The problem of how to determine the class of a given IP address was solved by encoding the class itself with the most significant bits of the address, as shown in Fig. 2.14: as this figure illustrates, the addresses of class A have the most significant bit

Table 2.1. IPv4 classful address ranges

Class	Address range	No of networks	No of addresses per network
A	0.0.0.0–127.255.255.255	$2^7 = 128$	$2^{24} = 16\,777\,216$
B	128.0.0.0–191.255.255.255	$2^{14} = 16\,384$	$2^{16} = 65\,536$
C	192.0.0.0–223.255.255.255	$2^{21} = 2\,097\,152$	$2^8 = 256$
D	224.0.0.0–239.255.255.255		
E	240.0.0.0–255.255.255.255		

set to 0, those of class B have the two most significant bits set to 10, those of class C have the three most significant bits set to 110, etc. So, if we wonder how to classify address 137.204.212.1, we quickly infer that it is a class B address, as the binary representation of 137 begins with 10; moreover, 137.204 is the net-ID and 212.1 is the host-ID. The address range for each class is detailed in Table 2.1, which also reports the number of IP networks (net-IDs) and the number of host addresses per network (host-IDs) in each class.

It is no wonder that the addressing scheme just described is known as *classful*.

Besides those included in classes D and E, additional IP addresses are reserved for special purposes and cannot be used to identify hosts uniquely. Some special cases, originally indicated in RFC 1700 [20], are:

0.0.0.0 – This very first class A address identifies "this host in this network" and is typically used as a source address when the sending host does not know its address yet. It can only be used as a source address.

0.x.y.z – This class A network is used as a source address to refer to "the specified host in this network": more accurately, given the notation for a generic IP address {net-ID, host-ID}, this special address is {0, host-ID}.

255.255.255.255 – This address identifies the so-called *limited broadcast*. It can only be used as a destination address and is employed as a destination address when the datagram has to be sent to every host connected to the same physical network as the source. A datagram with this address must never be forwarded outside the network the source belongs to. It is typically translated into a layer-2 broadcast address when the underlying physical network is a LAN.

127.x.y.z – This class A network is reserved to the *loopback* address. Typically, only the 127.0.0.1 value is employed, and it is used by a host to send datagrams to itself. It works even when there are no network interfaces installed.

Other reserved addresses are those where the host-ID bits are all set to zero or to one: the former identifies the whole IP network bearing the specified net-ID, the latter is the *directed broadcast* address, used as a destination address to reach all hosts of the IP network.

We can, therefore, say that the class B address 137.204.212.1 belongs to network 137.204.0.0, that a broadcast datagram to be received by every host in this network has to be sent to the destination 137.204.255.255, and that the addresses that can be assigned to the hosts fall in the 137.204.0.1–137.204.255.254 range. As Table 2.1 indicates, the number of addressable hosts in a class A, B or C network is $2^{24} - 2$, $2^{16} - 2$ and $2^{8} - 2$, respectively.

According to RFC 1918 [21], other addresses are reserved and confined in a *private IP address space*. Here the word *private* refers to enterprises and organizations employing the IP protocol to interconnect hosts, servers and routers inside their own network infrastructure, with no need to interact with *public* hosts and networks from the outside. The use of private networks is becoming more and more common for two reasons. First, if an organization is interested in interconnecting only its internal network resources, there is no need to apply for a public address space that has to be authorized and assigned by the specific administrative body through long and costly procedures. Second, for security reasons it is preferable to keep critical network resources isolated from the rest of the Internet, in order to prevent attacks from the outside that could compromise normal operations. And private addresses exactly allow one to do so, as they are not visible from the Internet! Every datagram that has a private destination or source address is rejected by routers in networks that do not use private addresses.

If an organization or home user wishes to set up a private IP network, it is allowed to choose from the following address space reserved by the IANA, the Internet Assigned Numbers Authority [22], for private use:

10.0.0.0–10.255.255.255,   a whole class A network;
172.16.0.0–172.31.255.255,   16 contiguous class B networks;
192.168.0.0–192.168.255.255,   256 contiguous class C networks.

Private IP addresses can be used without any coordination with the IANA. It is also possible to use the same address space on different private networks, as long as they are isolated from each other. For these reasons, the design and configuration of private IP networks is faster and more flexible than the use of public addresses. However, it is very likely that sooner or later the same organization or home user will want to connect its network to the public Internet. In this case, one solution is to apply for a public address space and reconfigure the hosts and routers that need to be connected to the outside. However, this approach could become burdensome; also, as we mentioned before, resources connected to the public network are potentially exposed to security threats. For this reason, an alternative solution based on the use of address translation techniques can be deployed, but we defer this topic to later chapters.

Originally, four billion addresses were deemed more than enough by the IPv4 developers. However, due to the reserved address ranges and mainly because of the classful allocation procedures used in the past, the 32-bit IPv4 address space is not considered sufficient any longer. On the other hand, the adoption of version 6 of the protocol, IPv6, providing a much larger address space, does not seem to be happening soon. Therefore, specific workarounds, such as classless addressing schemes, subnetting or supernetting techniques and private address-translation mechanisms have been and are currently being deployed as short- and mid-term solutions to overcome the lack of addresses. Let us examine some of them in detail.

2.4.2   Classless IP addressing

When the native, classful addressing scheme was still in use, the assignment of address space to organizations that had to be connected to the public Internet was done in a centralized manner, and according to the size of the requesting organization. It turned out that most of the organizations were too large for a class C network, so they were assigned a class B network even though their size was not large enough to require an entire class B space. This led to an inefficient address distribution, while the class B addressing space started to run out of available networks.

Back in the early 1990s, the need for a more flexible and efficient way to provide IPv4 addresses to end users led to the adoption of a classless addressing scheme and a hierarchical assignment procedure [23], which are still in use today. The IANA coordinates the numbering distribution. It allocates contiguous blocks of available addressing space to the five regional Internet registries (RIRs), AfriNIC for Africa, APNIC for Asia and Pacific, ARIN for North America, LACNIC for Latin America and some Caribbean Islands, RIPE NCC for Europe, the Middle East, and Central Asia. These RIRs are then responsible for allocating smaller address blocks to local Internet registries (LIRs) and Internet service providers (ISPs), which in turn assign some of these addresses to the end users.

However, all these allocations and assignments are not based on classes any longer: rather, they adopt the *classless inter-domain routing* (CIDR) standard. The idea is to define IP networks of suitable size by specifying the number of bits dedicated to the net-ID, that is not necessarily of one, two or three bytes, as in the classful scheme. Now a net-ID can be represented by any number of bits from 0 to 32, although networks with a net-ID shorter than eight bits have never been allocated [24]. According to the CIDR notation, IP networks are identified by a *prefix*, shown as the network address followed by the number of net-ID bits, the so-called *prefix length*. For instance, if the address 137.204.212.1 is part of a block of 2048 addresses, the corresponding network prefix is 137.204.208.0/21. To understand it, it is sufficient to look at the binary representation of the address,

$$10001001.11001100.11010100.00000001$$

and to observe that the host-ID is made of $\log_2 2048 = 11$ bits, while the net-ID length is 21 bits. Therefore, the network address is obtained by setting the 11 less significant bits to zero, i.e.,

$$10001001.11001100.11010000.00000000,$$

which corresponds to 137.204.208.0. A directed broadcast address is also defined by the CIDR standard as the address where all host-ID bits are set to one. In our example, it is the binary sequence

$$10001001.11001100.11010111.11111111,$$

Table 2.2. CIDR network prefixes

Network prefix	No. of addresses	Network prefix	No. of addresses
$N_8.0.0.0/8$	$2^{24}=16\,777\,216$	$N_8.N_8.N_5.0/21$	$2^{11}=2\,048$
$N_8.N_1.0.0/9$	$2^{23}=8\,388\,608$	$N_8.N_8.N_6.0/22$	$2^{10}=1\,024$
$N_8.N_2.0.0/10$	$2^{22}=4\,194\,304$	$N_8.N_8.N_7.0/23$	$2^9=512$
$N_8.N_3.0.0/11$	$2^{21}=2\,097\,152$	$N_8.N_8.N_8.0/24$	$2^8=256$
$N_8.N_4.0.0/12$	$2^{20}=1\,048\,576$	$N_8.N_8.N_8.N_1/25$	$2^7=128$
$N_8.N_5.0.0/13$	$2^{19}=524\,288$	$N_8.N_8.N_8.N_2/26$	$2^6=64$
$N_8.N_6.0.0/14$	$2^{18}=262\,144$	$N_8.N_8.N_8.N_3/27$	$2^5=32$
$N_8.N_7.0.0/15$	$2^{17}=131\,072$	$N_8.N_8.N_8.N_4/28$	$2^4=16$
$N_8.N_8.0.0/16$	$2^{16}=65\,536$	$N_8.N_8.N_8.N_5/29$	$2^3=8$
$N_8.N_8.N_1.0/17$	$2^{15}=32\,768$	$N_8.N_8.N_8.N_6/30$	$2^2=4$
$N_8.N_8.N_2.0/18$	$2^{14}=16\,384$	$N_8.N_8.N_8.N_7/31$	$2^1=2$
$N_8.N_8.N_3.0/19$	$2^{13}=8\,192$	$N_8.N_8.N_8.N_8/32$	$2^0=1$
$N_8.N_8.N_4.0/20$	$2^{12}=4\,096$		

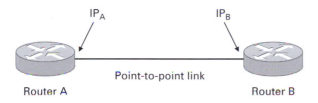

Fig. 2.15   Two routers interconnected by a point-to-point link

corresponding to 137.204.215.255. The remaining 2046 addresses between 137.204.208.1 and 137.204.215.254 can be used for host and router numbering.

Table 2.2 shows the general structure of all possible network prefixes with at least eight net-ID bits and the number of addresses in the corresponding block. In this table, N_i is an octet where only the i most significant bits may take on either the 1 or the 0 value, while the remaining $8-i$ bits are all set to zero, since they are part of the host-ID. Accordingly, N_8 represents an octet assuming all the possible values between 0 and 255. Former class A, B and C networks are represented by /8, /16 and /24 prefixes, respectively, while /32 prefixes are used to identify single IP addresses, typically as host-specific routes in routing tables employed by network devices. It might seem that /31 prefixes are of no use, since they include the network address and the directed broadcast only, leaving no space for usable addresses. We will shortly see that this is not always true, and that even these prefixes turn out to be handy.

The flexible use of the classless addressing scheme allows one to choose the IP network size that best fits the dimensions of the physical network to be addressed. An interesting case is represented by a point-to-point link that is typically used for router-to-router interconnection, as shown in Fig. 2.15, treated as a whole network at the IP layer. To comply with CIDR rules, /30 is the largest prefix that can be used in this case, since it provides four addresses, two reserved for network and directed broadcast, two to be used for the router interfaces. For instance, prefix 137.204.212.4/30 can be assigned to the point-to-point link, where 137.204.212.5 and 137.204.212.6 are the addresses assigned to the endpoint interfaces and 137.204.212.7 is the directed broadcast. However, this choice has the drawback of wasting half of the address space: to address N point-to-point links, N /30 prefixes are needed, for a total of $4N$ addresses, out of which only $2N$ are actually assigned to router interfaces.

To solve this problem, different solutions have been deployed by network administrators. One is to assign /30 prefixes taken from the private address space, but this requires a mixed use of public and private addresses that could cause configuration errors and routing problems. Another way is to completely avoid the assignment of IP addresses to point-to-point interfaces, based on the principle that addresses are not needed on a network segment where the sender can talk to a single receiver only. However, this approach leads to management problems, as unnumbered network segments are invisible to outside equipment. A third way is to assign a host-specific /32 prefix to each endpoint, but this is allowed only when the point-to-point protocol (PPP) encapsulation is used, as RFC 1661 [25] indicates. Furthermore, since the two addresses are not required to be contiguous, it may be impossible to identify the link as an IP network, leading to problems similar to those of unnumbered links. The last and more general solution, suggested by RFC 3021 [26], is to use /31 prefixes for point-to-point links where the only two possible addresses are assigned to the endpoint interfaces. For instance, if prefix 137.204.212.2/31 is assigned to the link, the two interfaces are addressed by 137.204.212.2 and 137.204.212.3. There is no need for a network address, since the prefix itself could be used to identify the link inside routing tables, while the limited broadcast 255.255.255.255 could be used instead of the missing directed broadcast.

2.4.3   Subnetting and supernetting

One of the main problems that network administrators have to deal with when designing an IP network is how to assign portions of the allocated address space to separate network segments. For instance, the network infrastructure of a big organization can be spread over several buildings and campuses, which are often very far one from the other. Not to mention the case of even bigger enterprise networks spanning over different cities or states. This problem was raised when the classful addressing scheme was used. Several organizations were assigned a class A or B network and they needed to split the address space among several physical network segments, either because of administrative issues or because of differences and limits imposed by the technology used in each segment.

The well-established solution to this problem is to adopt subnetting, i.e., to partition the address space logically into *subnets*, and to assign numbers from a given subnet to hosts connected to the same network segment [27]. Then, each subnet is treated as a separate object at the IP level and communications between different subnets are performed through layer-3 devices, such as routers. However, it must be pointed out that any information related to subnetting operations is relevant only to the equipment inside the current IP network, since its address space is seen as a whole block from the outside.

Table 2.3. Netmask octet values

Binary	Decimal
00000000	0
10000000	128
11000000	192
11100000	224
11110000	240
11111000	248
11111100	252
11111110	254
11111111	255

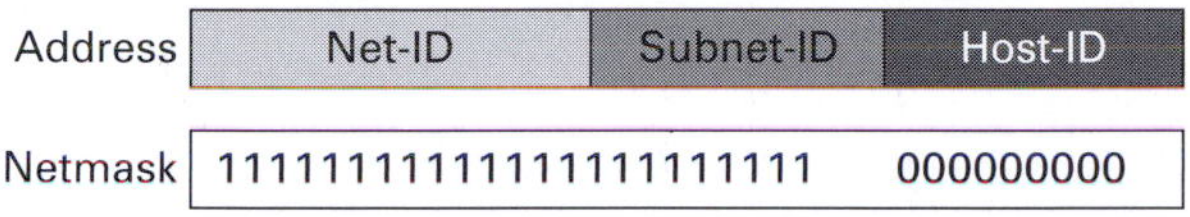

Fig. 2.16   Structure of an IPv4 address when using subnetting and related netmask

To specify the subnet, the IPv4 address structure has been modified as shown in Fig. 2.16, where some of the most significant bits of the previous host-ID are now dedicated to the *subnet-ID*, thus reducing the bits used for the actual host-ID. The number of subnet-ID bits to be used depends on the number of subnets required. For instance, if only two subnets are required, a single-bit subnet field is sufficient.

When an IP network has been subnetted, internal hosts and routers must be aware of how the address space has been partitioned. In other words, for each IP address, additional information on the number of subnet-ID bits is required. This is specified by the *netmask*, a 32-bit sequence where the bits corresponding to the net-ID and subnet-ID are set to one, while those dedicated to the host-ID are set to zero, as Fig. 2.16 exemplifies. Similarly to IP addresses, netmasks are typically specified in dotted decimal notation. Table 2.3 shows the nine possible values a netmask octet may assume and the corresponding decimal values. Netmasks corresponding to unpartitioned former class A, B and C networks were represented as 255.0.0.0, 255.255.0.0 and 255.255.255.0, respectively. When former class B networks were partitioned into 256 subnets, these were identified using the 255.255.255.0 netmask. In general, for each subnet the very first and last addresses are, as usual, reserved to the subnet address and to the directed broadcast. For instance, if network 137.204.0.0/16 is divided into 256 subnets, the address 137.204.212.1 with netmask 255.255.255.0 belongs to subnet 137.204.212.0 with directed broadcast 137.204.212.255. Since subnetting means partitioning the address space into 2^s subnets, where s is the number of subnet-ID bits, a network can only be divided in 2, 4, 8, 16, . . . non-overlapping subnets. A useful graphical representation of subnetting is displayed in Fig. 2.17, where the address space partitioning into fixed-size subnets is shown.

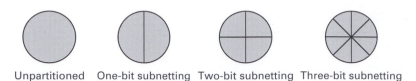

Unpartitioned   One-bit subnetting   Two-bit subnetting   Three-bit subnetting

Fig. 2.17   Graphical representation of fixed-size subnetting

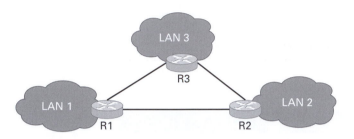

Fig. 2.18   Example of a network topology with three LANs and routers

However, it is very common that different network segments have different requirements in terms of the number of hosts to be addressed. In this case the use of fixed-size
subnetting is inefficient, as the subnet size suitable for the largest segment has also to
be employed for smaller segments, with a waste in address space. Therefore, subnets
of variable size can be resorted to, as long as their size is a power of two [28]. The
CIDR notation is again used in this case. As an example, let us consider the topology
in Fig. 2.18 representing the network infrastructure of an organization comprising three
LANs, the corresponding routers and the point-to-point connections between them. The
requirements in terms of number of addresses on each LAN are 126 for LAN 1 and
LAN 2 and 62 for LAN 3, to be used for host and router numbering within each LAN.
Six additional addresses are needed for the router-to-router interface numbering. The
total number of addresses required is 320, that is, nine bits are necessary. As a consequence, a /23 prefix has to be allocated: in the example, prefix 137.204.212.0/23 is
used. This address space is then partitioned into variable-size subnets – two /25 subnets
for LAN 1 and LAN 2, a /26 subnet for LAN 3 and three /31 subnets for point-to-
point links, as shown in Fig. 2.19. The corresponding address assignment is shown in
Table 2.4. The remaining address space from 137.204.213.70 to 137.204.213.255 is not
allocated and can be used for future subnet definition. For instance, if an additional network segment is added requiring 62 addresses, a subnet with prefix 137.204.213.128/26
can be assigned to it, as shown in Fig. 2.20. The only alternative choice is
137.204.213.192/26.

The introduction of the CIDR standard allows one to generalize the subnetting operation and to define a complementary *supernetting* procedure. While subnetting a network
prefix essentially means to increase the number of net-ID bits, the supernetting operation
consists of reducing it, while extending the host-ID field, as shown in Fig. 2.21. This
technique allows one to aggregate contiguous network prefixes and is mainly used to

Table 2.4. Solution with variable-size subnets

LAN	Subnet prefix	Netmask	Directed broadcast
LAN 1	137.204.212.0/25	255.255.255.128	137.204.212.127
LAN 2	137.204.212.128/25	255.255.255.128	137.204.212.255
LAN 3	137.204.213.0/26	255.255.255.192	137.204.213.63

PTP link	Subnet prefix	Netmask	Addresses
R1-R2	137.204.213.64/31	255.255.255.254	.213.64–.213.65
R2-R3	137.204.213.66/31	255.255.255.254	.213.66–.213.67
R1-R3	137.204.213.68/31	255.255.255.254	.213.68–.213.69

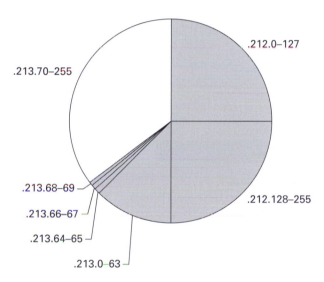

Fig. 2.19   Address space partitioning with variable-size subnets

reduce the number of entries in the routing tables of routing devices. For instance, the topology illustrated in Fig. 2.22 shows three networks with four prefixes assigned to them as well as the interconnecting routers. In general, router R5 must know a route for each of these prefixes, as it has to be able to route any packet directed to the three networks. According to the topology, R4 acts as R5's next hop and takes care of delivering the packets to R1, R2 or R3 depending on the destination address. However, since packets directed to the three networks are always forwarded from R5 to R4 through the same interface, if prefixes assigned to these networks are contiguous, it is possible to aggregate them in a single entry in the routing table of router R5. In our example, this supernetting operation is possible and the result is a single routing entry in R5 with prefix 137.204.208.0/21 and R4 as next hop. With similar operations the number of entries in the Internet core routers can be reduced, saving storage space and decreasing the table look-up time.

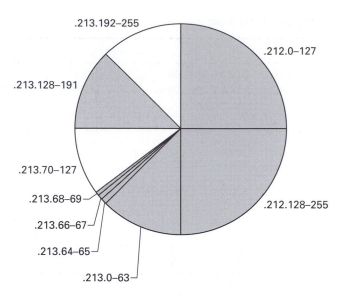

.213.192–255

.212.0–127

.213.128–191

.212.128–255

.213.70–127

.213.68–69

.213.66–67

.213.64–65

.213.0–63

Fig. 2.20   Address space partitioning with additional /26 subnet

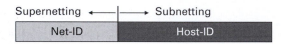

Supernetting ◄───────► Subnetting

Net-ID   Host-ID

Fig. 2.21   Duality of subnetting and supernetting operations

2.4.4   The delivery of IP packets

The previous sections introduced the concept of IP networks and subnets, which are separate entities interconnected by network-layer devices such as routers. The netmask concept was also presented as a way to specify the size of the network or subnet space a given IP address belongs to. This information is essential for a host, in order to perform the correct delivery of an IP packet. In particular, the sender is required to discriminate whether or not the destination address refers to a machine attached to the same network segment as the sender, i.e., whether or not source and destination addresses belong to the same IP network or subnet. If origin and destination are on the same network or subnet a *direct delivery* is performed, otherwise the sender goes for an *indirect delivery* and forwards the packet to a known intermediate node (a router), which takes care of either delivering the packet to the actual destination or forwarding it to another intermediate node.

In the example of Fig. 2.23, the host with address 137.204.212.18 and netmask 255.255.255.0 performs a direct delivery when sending an IP packet to 137.204.212.208: the target is on the same network and both source and destination addresses have the same net-ID, corresponding to prefix 137.204.212.0/24. On the other hand, when the same host wants to send a packet to 155.185.48.134, it finds out that applying its own netmask to this destination address results in a different net-ID, i.e. 155.185.48.0, so

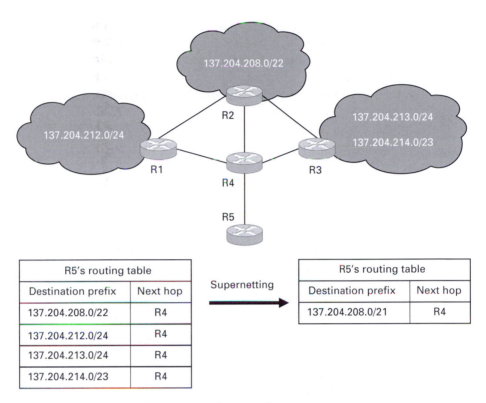

Fig. 2.22   Example of route aggregation by means of supernetting

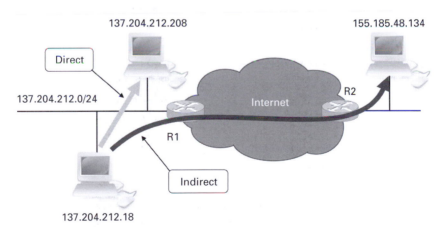

Fig. 2.23   Direct delivery vs. indirect delivery

an indirect delivery must be performed through router R1. The originating host must know something about R1 as a next hop, typically its IP address on the same network segment. Therefore, the netmask associated with the sender's IP address is used to check whether the destination address has the same net-ID, so that it is possible to discern the kind of delivery to be performed. A similar procedure is followed by intermediate routers when forwarding a packet: this however involves the look-up of the routing table, to be discussed in Chapters 6 and 7, devoted to routers, routing algorithms and protocols.

Once it has been decided whether to deliver a packet directly to the destination host or to forward it to an intermediate node, the actual transmission has to be carried out according to the specifications given by the physical and the data-link layer protocols used in the current network segment. This issue too will be discussed next, examining the case of Ethernet networks in Chapter 3.

2.5   TCP: transmission-control protocol

Now that the IP protocol has been described in detail, we conclude with a brief introduction of the two alternatives the TCP/IP protocol suite offers at the transport layer, TCP and UDP.

Focusing on TCP [29], we limit our description to its header format and its fundamental functionalities, aware of the fact that a thorough treatment of this protocol might well fill the pages of an entire book.

The TCP header is shown in Fig. 2.24, following the same convention adopted when presenting the IP header.

The first two bytes indicate the source port number, the following two bytes the destination port number. These values, together with the IP addresses of the sending and the receiving end, are extremely important: the {IP address, port number} pair represents the so-called *end point*, also known as the socket in Unix terminology; two end points

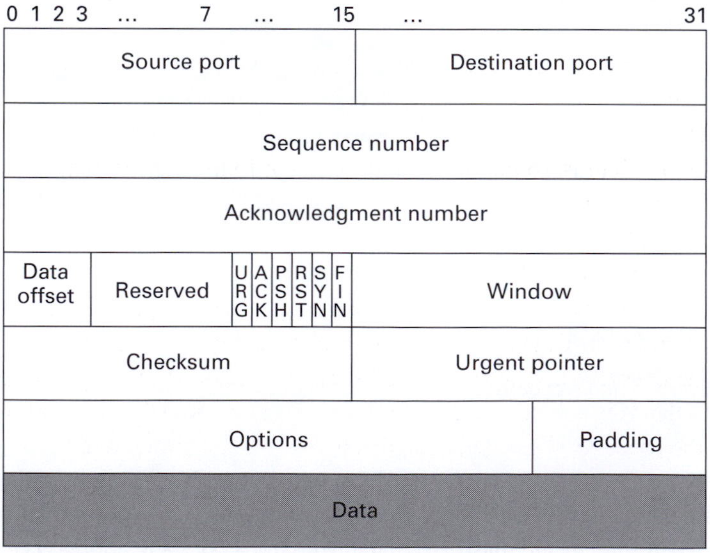

Fig. 2.24   TCP header

uniquely identify a TCP connection. So, if one end point is (155.185.52.156, 2048) and its counterpart is (129.78.64.24, 20), there is only one TCP connection over the Internet currently operating on such pairs of IP addresses and port numbers: equivalently, these end points tag, with no ambiguity, the sending process, i.e., the application that passed its bytes to TCP and the process – application – that has to receive those data. Note that in the TCP/IP protocol suite the end point is the equivalent of a layer-4 SAP in the OSI Model: through the end point an application can access the transport services TCP makes available.

To understand the meaning of the next field, let us state beforehand that the TCP numbers all bytes it sends, starting from an initial value, termed the initial sequence number (ISN), chosen during the opening phase of the TCP connection. This tagging operation is fundamental, as the TCP needs to identify all bytes, and if needed to retransmit the missing ones, to assure a reliable delivery service. The following field in the TCP header, 32 bits wide, termed the sequence number, therefore, reports the sequence number of the first byte of the payload within the segment: e.g., if the ISN chosen is 1618234416 and the sequence number is 1618234588, then the first byte of data is the 173rd of the original stream.

The acknowledgment number field, again 32 bits wide, contains the sequence number of the next byte of data that the TCP end is expecting to receive. As an example, if it contains the 1618239654 value, the TCP process is expecting byte 1618239654 (equivalently, it is acknowledging all bytes up to byte number 1618239653).

The data offset field, four bits wide, indicates the number of 32-bit words in the TCP header, which consists of an integer number of 32-bit words.

The six-bit reserved field is reserved for future use, and must be zero.

Six control bits or flags follow. URG, when set, indicates that the value of the urgent pointer field is significant. ACK, when set, indicates that the value of the acknowledgment field is significant. Then come the PSH, push function, and the RST, reset connection, which are used in special instances: the first to notify the need to immediately pass the data to the application above, the second to flush the connection. Next, two flag bits, SYN and FIN, that are set during the TCP connection opening and closing phase, respectively.

The window field is 16 bits wide, and bears the following meaning: it indicates the number of bytes that the sender of the TCP segment is willing to accept, beginning from the one indicated in the acknowledgment field.

The checksum field is 16 bits wide, and unlike the IP checksum, it covers both the header and the payload of the TCP segment. It is actually computed by prefixing a 96 bit pseudoheader, constructed as Fig. 2.25 indicates, to the TCP header: the pseudoheader includes the IP source and destination address, the indication of the protocol that passed the data to the TPC-sending module, and the length of the TPC segment, in bytes, not counting the pseudoheader. The value that appears in the urgent pointer field, when the URG flag is set, is an offset pointing to some data in the TCP segment, that it is urgent to pass to the receiving application as soon as possible.

Finally, the options field: of variable size, it encompasses the maximum segment size (MSS) option; this option is used exclusively during the opening connection phase and

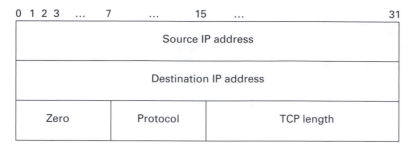

0 1 2 3 ... 7 ... 15 ... 31		
Source IP address		
Destination IP address		
Zero	Protocol	TCP length

Fig. 2.25   TCP pseudoheader

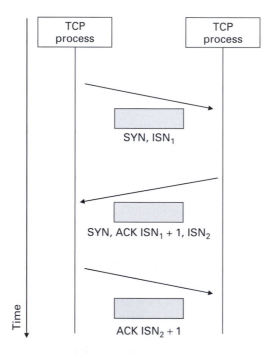

Fig. 2.26   TCP connection-opening phase

communicates the maximum segment size that the sending TCP process is willing to accept.

Padding field: as expected, a variable size, all-zero field, that guarantees the TCP header ends on a 32 bit boundary. It is not transmitted as part of the segment.

In what follows, we intentionally omit an accurate description of the TCP algorithms, and limit the treatment to its essential features.

First of all, let us recall that TCP is a connection-oriented protocol, meaning that a connection establishment phase and a connection termination phase are present. The first is depicted in Fig. 2.26, where time flows along the vertical axis and each arrow corresponds to the transmission of a TCP segment, exchanged between the two TCP ends. This figure represents the three-way handshake between the two TCP processes: no wonder this is the term traditionally employed to define the connection-opening procedure, as

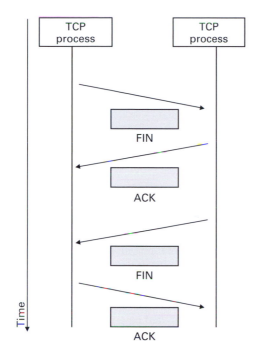

Fig. 2.27   TCP connection termination phase

three frames are actually exchanged. The first move is made by the TCP process that wants to open the connection, i.e., by the TCP process that is said to perform the active open, client side: no data are sent, only a segment exclusively constituted by its header, bearing the SYN flag set, the declaration of the initial sequence number (ISN)(do you remember that?) chosen and an indication of the maximum segment size (MSS). The TCP process at the other end replies, performing the so-called passive open, server side: it sends the second TCP segment in the figure, which acknowledges the first, and in turn declares the ISN and MSS for the opposite direction, as TCP guarantees a full-duplex type of connection. Its reception triggers the third segment, which acknowledges the former and opens the path to the actual data flow, which can now commence.

The orderly connection release is the one Fig. 2.27 describes, displaying the exchange of four TCP segments. Any of the two TCP ends can independently decide to shut down the connection: it sends a segment with the FIN flag set, and awaits for the corresponding ACK, carried by the second segment in the figure. The TCP process that begins the connection termination phase is said to perform the active close. In turn, the other direction is usually closed in an analogous manner, the second TCP end performing the passive close.

In between the opening and closing phases, data are exchanged between the two ends; the rate at which they are transmitted is shaped by several algorithms. The TCP employs the slow start, congestion avoidance, fast recovery and fast retransmit algorithms. Moreover, the most popular versions of TCP today adopt the SACK option [30], which allows the selective retransmission of only those bytes not acknowledged by the receiving end.

Fig. 2.28   UDP header

2.6   UDP: user datagram protocol

Everything that TCP has to offer, UDP does not possess. This protocol provides a bare transport functionality, as revealed by the presence of the source and destination port number fields, 16 bits wide as in TCP, and not much more, in the UDP header reported in Fig. 2.28.

The UDP header additionally includes: the length field, 16 bits wide, which indicates the length of the UDP datagram, including header and data, expressed in bytes; the checksum field, again, 16 bits in length, computed on a pseudoheader analogous to the one TCP employs, except for the UDP length replacing the TCP length, plus the UDP data field.

It is quite pleasing to cite the corresponding RFC, [31], as it consists of only three pages, references included!

Slim as it appears, UDP is often used as a transport solution over the Internet: besides video, voice and multimedia services, brief data exchanges of the one-shot request-reply type often rely on UDP: examples of these are the messages employed by the simple network management protocol (SNMP)[32], which allows one to monitor the device status in a network.

2.7   Exercises

2.1   Q – Is it correct to state that a class B network can be partitioned into 256 class C networks with a subnetting operation?

A – No, it is not correct. Class B and C addresses always differ for the second most significant bit. It is more correct to state that a class B network can be partitioned into 256 subnets that are equivalent to class C networks.

2.2   Q – Is it possible to use a public IP address space not allocated by IANA to number the hosts connected to an isolated network?

A – Yes, it is possible, as long as the network is isolated. However, in case some form of network translation is used to provide connectivity with the public Internet, the use of unallocated public addresses within the internal network could compromise the correct communication when the same address space is eventually allocated.

2.3 **Q** – Which of the following are correct network prefixes?

(a) 137.204.15.128/25;
(b) 7.16.129.0/17;
(c) 201.129.80.0/20;
(d) 90.0.0.0/7;
(e) 12.4.0.128/24;
(f) 172.16.230.154/30;
(g) 77.14.228.112/28;
(h) 10.55.28.3/31.

A – The correct prefixes are those whose binary representation shows all zeros in the host-ID, i.e. (a), (c), (d) and (g).

2.4 **Q** – What network does the host with IP address 137.205.211.141/25 belong to?

A – In hybrid notation, the host address is 137.205.211.10001101/25: the first 25 bits identify the network, i.e., 137.205.211.10000000/25, whose address in dotted decimal notation is 137.205.211.128/25.

2.5 **Q** – Organization Alpha has been assigned by its registry the block of addresses 137.205.40.0/21. How many hosts can the organization have?

A – The organization has a space for assigning IDs of $32 - 21 = 11$ bits. It can therefore have $2^{11} - 2 = 2046$ hosts, whose addresses range from 137.205.40.1/21 to 137.205.47.254/21.

2.6 **Q** – Organization Beta has been assigned the block of addresses 137.205.41.128/25. What are the smallest and the longest subnets that can be used by Beta?

A – The smallest subnet is identified by the /31 prefix, the longest by the /26 prefix.

2.7 **Q** – Consider the example in Fig. 2.18 and the related address plan in Fig. 2.19. Consider also that an additional network segment requiring 62 addresses must be allocated. Why can only subnet prefix 137.204.213.128/26 or 137.204.213.192/26 be assigned? Why cannot 137.204.213.70/26 be considered as an option?

A – Since the additional segment requires 62 addresses, it must be assigned a valid /26 subnet prefix in the available address space. Valid /26 prefixes are those with the six least significant bits set to zero, i.e. 137.204.213.128 and 137.204.213.192 in the available space. 137.204.213.70/26 is not a valid /26 prefix.

2.8 **Q** – Which of the following prefixes are correct results of supernetting operations aggregating networks 195.185.56.0/21, 195.185.52.0/23 and 195.185.49.0/24 among others?

(a) 195.185.49.0/20;
(b) 195.185.32.0/19;
(c) 195.185.48.0/21;
(d) 195.185.48.0/20.

A – (b) and (d) are correct. (a) is wrong since it is not a valid /20 prefix. (c) is wrong because it does not include prefix 195.185.56.0/21.

2.9 **Q** – Assume that network 194.17.78.0/24 has been partitioned into two /25 subnets assigned to two separate network segments interconnected by a router. Assume also that a host connected to subnet 194.17.78.128/25 has been configured with address 194.17.78.204 and netmask 255.255.255.0. What happens when such host must send a packet to destination 194.17.78.3? And to 194.17.78.199?

A – Since the host has been configured with a wrong netmask, it believes that the entire network 194.17.78.0/24 is attached to the same segment. Therefore, it tries to perform a direct delivery for every packet destined to an address within this network. In particular, it fails to send the packet to 194.17.78.3, which is attached to the other segment, while it is still able to communicate with 194.17.78.199.

3   Ethernet networks

The initial focus of this chapter is on the most significant issue to solve in the MAC sublayer of LANs and MANs: how to allow a given number of stations to share a common channel to communicate. Among the different solutions that can be devised, the answer successfully adopted by current LANs is introduced and critically commented on.

Such treatment is followed by a description of the general IEEE framework where LAN and MAN standards sit. There finds its place the standard for the implementation of Ethernet LANs, also known as the IEEE 802.3 standard, whose physical layer and MAC sublayer specifications are faced next. An adequate coverage is provided for the admissible topologies, MAC frame format and main algorithms; the evolution that this type of LAN has undergone in the last years is also illustrated, from the original solutions to fast and current gigabit Ethernet networks.

What follows is a description of the reference standards that modern cabling systems have to conform to. The intent is twofold: to make the reader understand (i) how crucial the cabling system is in guaranteeing a successfully operative network; (ii) how important its physical properties are when striving to support higher and higher data rates.

The chapter is closed by a few dense practices, which aim at explaining several real-world issues: the relation between layer-3 and layer-2 addressing schemes; the configuration of the hardware device that allows the station to connect to the LAN and to the Internet; and the layout of the cabling system for a campus network.

3.1   Multiple access

Before plunging into the nitty-gritty details of the standards that specify how to implement local area networks, it is useful to frame from a historical and logical perspective the conceptually juicy problem that needs to be faced in this type of networks. The last statement points to the more general issue of accessing a common communication channel, shared by a given number of stations, say M, that intend to exchange information through it, therefore clustering in, e.g., a LAN. The question that awaits an answer is: how do we coordinate the stations – given we decide to do so – and somehow discipline their transmissions so as to allow a profitable data exchange?

There are many options available when confronting general theme, faced in LANs, MANs and, to a more limited extent, in WANs, and different categories of channel access strategies can be delineated.

At one extreme of the scale sits the category of fixed channel access schemes: within this class, the resources available for communication are statically partitioned between the stations, a rigid approach well embodied by the frequency division multiple access (FDMA) and time division multiple access (TDMA) strategies, the main actors within this class.

In FDMA, the shared resource for communication is represented by a frequency band: in a first, simplified description of this access technique, the band is statically partitioned into M sub-bands, and each station can exclusively transmit on the sub-band preliminarily assigned to it.

In TDMA, the resource to be distributed among the M stations is time: the time axis is ideally divided into TDMA frames, a frame being constituted by M contiguous time slots. The TDMA frames repeat, one after the other, along the time axis. In each frame, a station is allowed to transmit only during the time slot assigned to it. Note that in this context the term "frame" does not have to be confused with the same word employed to indicate the MAC sublayer data unit: here it simply identifies a time interval of finite, adequate length.

Roughly, for both access solutions it can be stated that a station statically employs a fraction, $1/M$ when an equal share is required, of the total system resources to transmit its data. It will be a frequency sub-band in FDMA, a time slot in each frame for TDMA.

There is a major drawback inherently tied with static resource assignment: each station is allotted a fixed portion of the communication channel, regardless of its current transmission needs, which can dynamically vary and depend upon the traffic intensity that the station exhibits in different time intervals. What might happen with fixed channel assignment is that while a station sits idle, others can have data that need to be transmitted, forced to wait unnecessarily in the station queue due to lack of resources. Definitely not the best solution for data sources, featuring a significant value of the burstiness factor!

At the other extreme of the scale, one could opt for an intentional, absolute lack of coordination among stations. When a station has some data – MAC frame – to transmit, it immediately grabs the common communication channel and proceeds, regardless of what other stations might be doing.

It is not a surprise that the first access technique proposed and successfully adopted within the random access category obeys this philosophy. Its fascinating name is Aloha, stemming from the University of Hawaii, where this strategy was conceived and adopted to create the first packet-switched – satellite – network. As Fig. 3.1 portrays, when the Aloha random access technique is employed, stations inject their frames onto the common communication channel as soon as they are ready to be transmitted. Indeed, it is quite likely that frames transmitted by different stations will overlap, causing a collision phenomenon, whose frequency of occurrence will depend on the intensity of the traffic loading the network. A straight consequence is the loss of the frames involved in the collision, the need of frequent retransmissions and ultimately a low channel utilization: this is the price to be paid for the simplicity of the approach, where stations share the communication channel in a totally decentralized, uncoordinated manner.

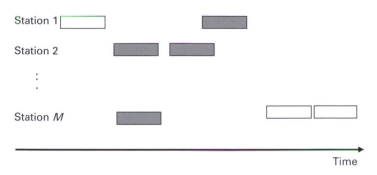

Station 1

Station 2

⋮

Station *M*

Time

Fig. 3.1   Typical frame sequences in an Aloha network (shaded boxes represent colliding frames)

The original Aloha and the numerous children this technique has spurred frequently appear in scientific literature and also find interesting applications in modern networking standards: their most prominent examples and their performance evaluation are covered in milestone books, such as [1] and [2], that although aged, are worth being pointed out as excellent starting-point references.

Within the class of random access techniques, a subfamily of strategies that is of interest to understand the MAC sublayer of current LANs, hides behind an acronym, CSMA, which stands for carrier sense multiple access. A strategy of this type puts into practice the first, intuitive countermeasure that comes to mind when the goal is to reduce collisions and improve channel efficiency with respect to Aloha: why not sense the shared communication channel before transmitting? A station can first figure out the channel status, detecting the presence of activity, then transmit a frame or postpone such transmission, basing its decision on the sensing result. The transmission attempt should definitely be deferred if the channel is busy. Simple as it sounds, this approach guarantees a significant throughput improvement, retains an appealing decentralized feature and introduces a modest implementation complexity.

In passing, note that the channel access schemes falling within the random access category are also termed contention-based strategies, a definition that originates from their most important feature: stations compete to grab the resources, with no field judge policing their behavior.

Before commenting on CSMA, its flavors and addenda in greater detail, let us complete the picture. In between static resource assignment techniques and random, totally uncoordinated access solutions, there lies the class of dynamic control. In this group, the resources available to communicate are assigned on demand, on the basis of the actual needs each station raises: inevitably, a tight coordination among stations, either in a distributed or centralized fashion, is required. This implies a potentially optimal deployment of the resources, but also suggests that the price to be paid is a non-negligible control overhead and, ultimately, an increased complexity.

The simplest setting to understand the on-demand resource assignment approach is well represented by a network featuring a star topology, where a master device at the star center coordinates the stations: the master periodically polls the stations to know

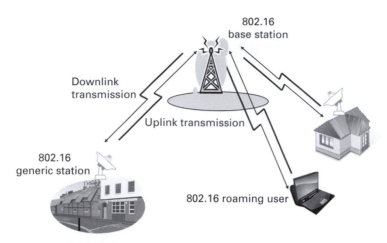

Fig. 3.2   IEEE 802.16 reference setting for Wireless MANs

their transmission requirements, then it assigns each of them an adequate portion of the shared resources to communicate.

To make the example more realistic, tying the generic description to the actual specifications of the IEEE standard for Wireless MAN, reference is made to the IEEE 802.16 documents and to the setting that Fig. 3.2 illustrates. In terms of channel access, let us emphasize that the wireless stations cannot directly communicate: rather, they rely on the central base station and bidirectional communications – from the station to the base and from the base to the station – are supported via, e.g., a frequency division duplex (FDD) approach. As the base station is the only user willing to transmit on the frequency sub-band dedicated to the downlink (from the base to the wireless stations) communications, no channel access issue actually exists on this subchannel. Its frequency sub-band is utilized logically, structuring data into frames (again, proper time units) of adequate length. In each frame, the time division multiplex technique allows the separation of data intended for different wireless stations: in terms of control, the only action the base is required to perform is to broadcast through an appropriate downlink map (DL-Map) when each station has to lock onto this downlink transmissions to extract its data.

Things are different when the uplink (from the wireless stations to the base) communications are considered: the stations share a common frequency sub-band to send their packets to the base and some form of coordination is indeed required: here the dynamic control access technique comes into play. A forcedly primitive summary of the IEEE 802.16 MAC sublayer indicates that stations inform the base of their transmission needs, periodically transmitting their new requests or upgrades on the uplink channel during some time slots specifically reserved to this task. Given the available bandwidth on the uplink sub-frequency band, the base returns the stations a temporal uplink map (UL-Map) specifying when and for how long each station holds the permission to transmit: the corresponding intervals do not overlap and the risk of collisions is, a priori, avoided. Figure 3.3 anticipates the salient details of the downlink subframe that the base station periodically broadcasts on the downlink: its second field, termed broadcast control,

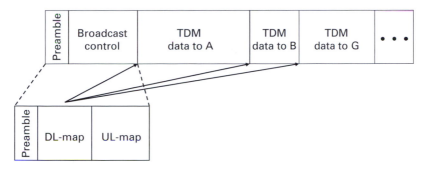

Fig. 3.3   A schematic representation of the DL subframe for the TDD IEEE 802.16 standard

provides the UL and DL-maps that inform the stations of the resource assignment decisions of the base. In this example, the subsequent fields contain data whose recipients
are in turn stations A, B and G.

It is interesting to note that these decisions depend not only on the amount of traffic
raised by the stations, but also on the type of traffic considered: different priorities are
assigned to real-time and non-real-time traffic, scaling the corresponding requirements
in four different MAC service classes. There will be time in Chapter 4 to pick up the
IEEE 802.16 standard again, with an accurate, critical treatment.

But let us go back to the CSMA strategies, selectively describing some of them.

3.1.1   Carrier sense multiple access strategies

As anticipated before, any random access technique belonging to the carrier sense family
displays the following feature: the station that has a frame ready to be transmitted over
the common access channel verifies whether some activity is present on it, indicating
that another station is already engaged in a communication. If this is the case, the packet
transmission is deferred. Depending on the station behavior after the sensing operation,
different CSMA techniques can be devised. A few examples are worth mentioning: first,
the *1-persistent* CSMA approach, where the station senses the channel and if it detects
it is idle, immediately transmits; if the channel is sensed as busy, the station holds back,
but ceaselessly monitors the channel activity: as soon as the idle state is detected, the
station places its frame on the channel.

Then, the *non-persistent* CSMA: as before, if the channel is sensed as busy, the station
defers its transmission; yet the station does not persistently hook onto the channel waiting
for its idle state. Rather, it delays the sensing operation for a random time, behaving in a
less aggressive manner than the *one-persistent* approach. If the channel is sensed as idle,
the station implementing *non-persistent* CSMA immediately proceeds with the packet
transmission.

A still different approach is displayed by the *p-persistent* CSMA version: now if the
channel is busy, the station does not transmit but keeps monitoring the channel; if the
channel is idle, the station transmits with probability p and with probability $1 - p$ delays

its transmission for at least τ seconds, where τ is the maximum time it takes for the signal to propagate between the interfaces of the furthest pair of stations sharing the channel. If readers are wondering why exactly τ seconds, we just ask them to hold their breath until the very next paragraph, which provides the necessary explanations. As for p, it is a parameter whose value can be varied, with the aim of optimizing the throughput-delay performance of the algorithm.

Let us first ask if the collision phenomenon disappears when CSMA is considered: the immediate answer is no, it does not. It is easy to portray an example to sustain this statement: two stations that implement either *1-persistent* or *non-persistent* CSMA simultaneously sense an idle channel, both correctly conclude that no one else is at the moment transmitting and, therefore, transmit their frames onto the common resource, heading toward an inevitable collision. More accurately, the uncertainty about the channel status is related to the time it takes the signal to propagate from one station interface to another: if sensing is performed *before* the signal already occupying the common channel has arrived at the station interface, this leads the station to erroneously conclude that yes, the channel is idle! In any such circumstance collision is inevitable.

It is, therefore, clear that τ, the propagation time between the two stations that are the furthest away, is a significant indicator: its value points to the worst case that can be encountered when assessing the achievable performance of any CSMA strategy.

So, going back to the *p-persistent* CSMA philosophy (when the channel is idle, transmission is intentionally delayed with probability $1 - p$ for at least τ seconds), the reader should observe that this choice aims to minimize the risk of collisions.

Even without resorting to quantitative performance evaluation, there should be no surprise behind the statement that the propagation delay τ plays a major role when assessing the behavior of any algorithm within the CSMA class. Letting intuition drive the reasoning, it is τ normalized by the time t_I it takes to transmit all bits in a MAC frame to shape both the behavior of the traffic-throughput curves and the relation between the throughput and the delay incurred by the frame before reaching its – local – destination. For those who are interested in a quantitative comparison of the performance of different CSMA techniques, the original studies in references [33] and [34] represent a good starting point.

As for the procedure to handle the inevitable retransmissions that a few packets will incur, the rationale behind it is that some random delay has to be forced between successive retransmissions: this helps in reducing the risk of further collisions.

The last remark of this subsection is devoted to a countermeasure that can be usefully combined with CSMA. Given that collisions will nevertheless be present, why not require each station to keep monitoring the channel even after the packet transmission has commenced? This allows the stations to detect collisions, revealed by the simultaneous transmission and reception of signals at the station interface, and therefore immediately stop the transmission of the corresponding frames so as to free the channel. This counteraction allows some layer-2 bandwidth to be saved and goes by the name of collision detection (CD). Its introduction leads to a considerable improvement in terms of achievable throughput and transfer delay and it is profitably implemented in conjunction with CSMA.

3.2   IEEE 802.3 and the IEEE 802 project

In the past, quite a few solutions to implement a LAN have been standardized. All of them come from the IEEE, which developed an entire framework, the so-called IEEE 802 project, where the standards for local and metropolitan area networks fit. The logical organization of the IEEE 802 project is summarized in Fig. 3.4, which reports the entire family of standards, whose specifications deal with the physical and data-link layers as defined by the OSI model. In the IEEE tradition, each standard is identified by a number: we will not cite all of them; rather, we will confine our attention to those in Fig. 3.4 that are worth mentioning when dealing with the LAN and MAN context. To this regard, the 802.3, 802.4 and 802.5 standard documents provide the specifications to develop three different – and totally incompatible – LAN solutions; the 802.6 standard was devised for MANs (although it did not definitely experience a great success); the 802.11 standard is for wireless LANs and the 802.16 for wireless MANs.

Restricting the focus to LANs only, the so-called 802.4, *token bus* standard, never gained any significant popularity. A different destiny was expected of the IEEE 802.5 standard, whose access method to the common channel well fits the on-demand category section introduced previously. Its MAC sublayer access solution, based on a decentralized approach, relies on a logical ring topology: stations overlook the ring, each having the possibility to transmit and receive from it via a proper interface. The 802.5 MAC dictates that only one station at a time holds permission to transmit its frames over the ring, thus eliminating the risk of collisions; the grant for transmission is represented by a special control frame, termed token, which in the absence of activity, keeps circulating over the ring, sequentially traversing all station interfaces. The essence of the access method can be described in the following manner: as soon as one station has one or more frames to transmit, it waits until the token arrives at its interface, it removes it from the ring – now it has permission to transmit! – and inserts its frames onto the ring; the frames will sequentially go through the interfaces of all stations, giving the actual destination the possibility of reading them. When these frames show up again at the sending interface

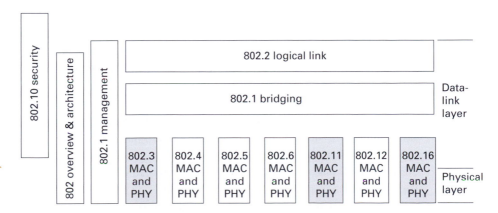

Fig. 3.4   The IEEE 802 project

they will be removed, and a new token will be released. The next downstream station waiting for transmission can capture it and in turn proceed with the insertion of its frames onto the ring.

Conceptually clean as this access algorithm might sound, it hides several non-trivial issues that the IEEE 802.5 MAC sublayer had to solve: undoubtedly the ring has to be managed, to cope with undesired phenomena such as the loss of the token or multiple tokens that erroneously circulate over the ring, as well as with ring failures, to name just the most evident anomalies that may occur. Some form of coordination and control is, therefore, needed: a monitor station is required, equipped with numerous timers, whose timing out reveals some undesired behavior and calls for adequate countermeasures. In 802.5 any station can potentially be the monitor, but only one has been elected to this role, in the LAN initialization phase.

These 802.5 LANs were indeed installed in the past: guaranteeing excellent performance, they however exhibited a sophisticated control and management plane. In the long run, this inevitably decreed their disappearance, in favor of the dominant 802.3 solution, astonishingly simple, yet highly robust. On the standard side, the IEEE 802.5 specifications were frozen in 2002.

The IEEE 802.3 standard, also popularly known as Ethernet, is the pick of the bunch. What is the main feature of its MAC sublayer? To rule the access of the stations to the shared communication channel via the *1-persistent* CSMA/CD algorithm. Now the simplicity is explained: neither management nor control is required by this random access approach, totally distributed and with no coordination among different stations.

Where does the more popular denomination Ethernet come from? The first studies that led to the standard were performed at Xerox, and the corresponding local area network was called Ethernet, from ether, a term indicating the common channel – a wired bus – that linked the stations. The project evolved in time, gaining momentum from the alliance between Xerox, Digital and Intel and led to the first core of the IEEE standard. Adhering to history and to common usage, in what follows we will interchangeably use the terms 802.3 LAN and Ethernet network.

But let us proceed in covering the essentials about the IEEE 802.3 requirements at the two bottom layers of the OSI model.

3.2.1   Reference topologies

In the old days the simplest topology exhibited by Ethernet networks was the bus, originally made of a segment of coaxial cable, suitably terminated at both ends to avoid signal reflections. The bus, exemplified in Fig. 3.5, is the communication channel shared by all stations; each station is in turn equipped with a network interface card (NIC), a hardware unit providing LAN connectivity, hence implementing the CSMA/CD access method, as well as other functionalities.

Note that all stations connected to the bus belong to what is called the same *collision domain*, as all their frames can potentially collide.

The first standard documents specified tight requirements regarding, e.g., the maximum number of stations allowed on the cable and the maximum cable length. These

Fig. 3.5   The simple bus topology

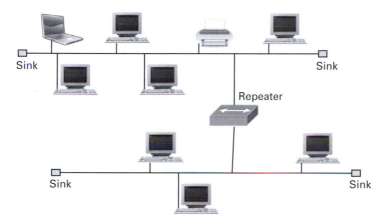

Fig. 3.6   Two segments interconnected via a repeater

limitations could be overcome, therefore connecting more stations to the same LAN and covering greater distances, by the introduction of a simple interconnecting device, called a repeater. Its functionalities exclusively reside within the physical layer (layer-1) and its job can be summarized as follows: a repeater amplifies and forwards the signal that it receives onto one of its ports toward all other ports. Fig. 3.6 schematically reports the case of a repeater that interconnects two 802.3 LAN segments. From a logical view-point, the repeater does nothing but widen the collision domain that the single segment features: good for the purposes mentioned above, ultimately not that exciting in terms of performance. As a matter of fact, if the number of stations connected to the LAN increases, collisions will become more frequent, and for high traffic levels performance will inevitably degrade.

In current implementations, the bus topology has been abandoned, in favor of the star portrayed in Fig. 3.7: at its center there appears a special device, which can be either a multiport repeater, also called a hub, or a switch, with completely different characteristics. In what follows, we will initially stick to the first case, observing that here too collisions occur: they correspond to two or more frames simultaneously reaching the repeater ports. Now it is the repetear that has to detect the collision and inform the stations.

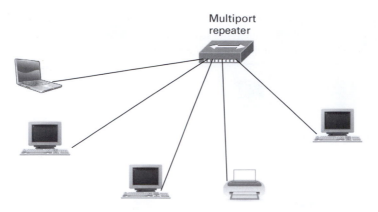

Fig. 3.7   A star with a hub at its center

Fig. 3.8   A direct, point-to-point connection in 802.3

There is still another reference topology for Ethernet, the one represented in Fig. 3.8: this scenario describes a plain point-to-point type of connection, where, e.g., a single station communicates via 802.3 to a switch.

Both the star and the point-to-point topologies represent the initial building blocks of Ethernet LANs, as they allow the construction of more complex and articulated infrastructures than those presented so far.

3.2.2   MAC sublayer

Modern terminology referring to Ethernet networks distinguishes between half-duplex and full-duplex transmission: mentioning the former inevitably implies the contention-based access method based on the *1-persistent* CSMA/CD access technique; on the contrary, the latter term points to an option that the 802.3 MAC provides, i.e., to allow simultaneous two-way transmissions over a point-to-point link. Let us describe them both.

3.2.2.1   Half-duplex transmission

When there is a true need to share a channel, as in the examples provided by the bus network and by the star solution with a hub at its center, the half-duplex transmission is a necessity and CSMA/CD is the solution. So, if the channel is sensed to be idle, the station proceeds with transmission (after a brief interframe gap); if the channel is sensed to be busy, the station holds back and keeps sensing it, postponing the transmission until the channel becomes idle again. Once a collision occurs, as soon as the station detects it, it immediately stops the frame transmission, and transmits a jamming sequence onto the LAN, to notify other stations that a collision took place and that some "garbage" is actually circulating over the LAN. When repeaters are employed, it is the repeater that detects the collision and transmits the jamming sequence.

After a collision, retransmissions have to be scheduled. In the IEEE 802.3 standard, the half-duplex operating mode details a controlled randomization process called "truncated binary exponential back-off." This algorithm mandates that, after detecting the collision, the MAC sublayer of the stations whose frames collided introduce some delay before attempting to retransmit. The delay is an integer multiple of an interval, termed slot time, equal to 512 bit times for 802.3 LANs operating at 10 and 100 Mbit/s, and to 4096 bit times for 1 Gbit/s 802.3 LANs.

Let us describe and comment on it in greater detail. Before attempting to retransmit for the first time, the station chooses with equal probability to wait either 0 or 1 time slot; if a second collision occurs, then the station waits for a uniformly distributed number of time slots in the range $[0, 3]$. For the kth retransmission attempt, $k \leq 10$, the number of slots the station delays the retransmission attempt is chosen as a uniformly distributed random integer in the $[0, 2^{k-1}]$ range. For $10 \leq k \leq 16$, the randomization range is frozen to $[0, 2^{10}]$, and if the limit to the maximum number of retransmission attempts is reached, that is 16, this event is reported as an error.

The main feature of this algorithm is to adapt the temporal window where retransmission attempts are distributed to the different traffic conditions of the 802.3 LAN. When traffic is moderate, it is quite likely that the first or, at most, the second retransmission attempt will be successful: if this is the case, the retransmitted frame will incur a modest delay. In high traffic conditions, more consecutive collisions may occur: now the randomization window is gradually widened, with the intent of reducing the probability of more collisions; the inevitable drawback is the increased delay that the frame experiences before reaching its destination.

3.2.2.2 Full-duplex transmission

In the much simpler situation of Fig. 3.8, if:

- The underlying physical media support full-duplex operation;
- Both stations at the end of the point-to-point link can simultaneously transmit and receive;

then there is no need to contend, and transmission can commence as soon as there are frames ready to be transmitted. What happens is that the CSMA/CD functionalities of the NICs are turned off and frames are sent over the link back-to-back, with a small interframe gap in between.

In this context, the only conceptually significant problem that might arise and that awaits a solution is to tune the two stations, exerting flow control: when the receiving

7	1	6	6	2	0-1500	0-46	4
Preamble	SFD	Dest ADD	Source ADD	Length/ type	Data	PAD	CHKSUM

Fig. 3.9   Frame format in IEEE 802.3

station at one end of the link is slow, it has to be able to notify the sending station of its near-congested status and to ask it to defer from sending more frames for some time.

3.2.2.3   IEEE 802.3 frame format

Before listing the actual Ethernet specifications for its different 10, 100 and 1000 Mbit/s versions, let us proceed in commenting on the frame format the standard defines, applying to all Ethernet networks. It is reported in Fig. 3.9, which follows the usual notation: the least significant bits, also the ones to be transmitted first onto the LAN, appear in the leftmost position. The size of each field in the frame, expressed in bytes, appears on the upper part of the figure: all these fields have a fixed size, except for the data and pad, for which the size range is provided. The first field is the preamble: it is seven bytes long and the pattern of each byte is 10101010. Its known sequence allows the physical layer circuitry of the receiver to synchronize with the frame's timing.

Then comes the start of frame delimiter (SFD): it is one byte, 10101011, and indicates the start of the frame.

Next, there appear the addresses of the frame's source and destination: the destination address field specifies the destination address for which the frame is intended; the source address field identifies the station sending the frame. They are typically referenced as MAC addresses, hardware addresses or physical addresses, and are six bytes (48 bits) long. A MAC address is traditionally provided in hexadecimal notation, bytes being separated by colons: an example would be 8:0:20:12:5A:4A.

The first, least significant bit of any 802.3 MAC address is termed the I/G bit, which stands for individual/group, and bears a meaning only for the destination address: when set to 0, it indicates a single, unicast destination, whereas set to 1, it indicates that the destination address field contains a group address, typically identifying a group of stations connected to the LAN; in the source address field, this first bit is reserved and set to 0.

The second bit of the MAC address allows distinction between locally or globally administered addresses. For globally administered addresses (termed U, universal, by the standard), the bit is set to 0. Vice versa, if an address is to be assigned locally, this bit is set to 1. Although the standard provides this option, the universal addressing scheme is the one that current 802.3 LANs adopt. Put into practice, this means that any 802.3 NIC comes shipped by the manufacturer with its MAC universal address written in its ROM: there is no other 802.3 NIC in the world identified by the same address!

Coordination in assigning universal addresses is guaranteed by the IEEE itself, which manages the MAC address space: a manufacturer of 802.3 NICs can purchase addresses in blocks (the minimum block size is 2^{24}), paying a fee to the IEEE. The structure of the MAC universal address reflects the description above: the first three bytes of the address, 8:0:20 in the previous example, provide the organization unique identifier (OUI), i.e., the identity of the manufacturer; the remaining three bytes identify the specific NIC made by that company, NIC number 12:5A:4A in the example. A practical fact worth noting is that at the URL, http://standards.ieee.org/regauth/oui/index.shtml, the IEEE registration authority allows one to search for the manufacturer corresponding to a known OUI; the entire database that sets the OUI–manufacturer correspondence is downloadable, and it is used by several network monitoring tools.

As an aside, we observe that although MAC addresses are assigned to the hardware by the manufacturer and in principle a NIC is not supposed to modify its MAC address, recent operating systems provide tools that allow privileged users to force a NIC to act as if it was assigned a different MAC address, specified by the user. This operation is called MAC spoofing and is often used to elude MAC-filtering policies and to perpetrate those LAN-based security attacks known as man-in-the-middle attacks.

One last observation regarding addresses is reserved for the MAC broadcast address: used as a destination, it allows one to send a frame to the totality of the LAN stations. Its standard value is FF:FF:FF:FF:FF:FF in hexadecimal notation.

As for the length/type field, it is two bytes long and its meaning depends on the numeric value it takes. If this value is lower than 1500, (the maximum size in bytes allowed for the data field is exactly 1500), then this field indicates the length of the data field. If its value is higher than 1500, then this field provides type information, indicating the network protocol that passed its data to the MAC sublayer when the frame was created at the sending station. In current implementations, last usage prevails, and as the reference protocol at network layer is IP, it is the corresponding value, 08-00 in hexadecimal notation, to appear in this field.

The payload encapsulated in the 802.3 MAC frame follows: the data field can be of variable length, ranging from 0 to 1500 bytes. However, as a minimum frame size is required for correct CSMA/CD protocol operation, when necessary the data field is extended by appending some extra bits through the following field, termed pad: pad bits carry no actual meaning, they simply ensure that the frame reaches the minimum size needed, equal to 64 bytes (512 bits) counted from the source address field to the end of the frame.

The requirement of the minimum size imposed to any 802.3 frame guarantees that all collisions are detected. To understand it, let us examine the original bus topology and consider two stations, X and Y, under the hypothesis that they are the furthest pair of stations in the LAN. Imagine that at time t_0 station X senses the channel, which is indeed idle: as a consequence X transmits its frame. Recalling that τ indicates the maximum propagation delay, let station Y have a frame to transmit right before time $t_0 + \tau$: Y erroneously concludes that the channel is idle, hence transmits its frame. Station Y will almost immediately learn about the collision, but what about station X? The standard requires that the frame from station X be long enough so that X is still sending it at time

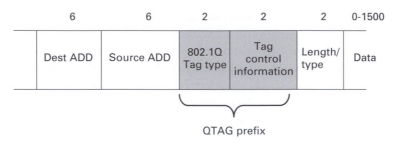

6	6	2	2	2	0-1500
Dest ADD	Source ADD	802.1Q Tag type	Tag control information	Length/ type	Data

QTAG prefix

Fig. 3.10   The additional VLAN header

$t + 2\tau$; if it weren't so, the NIC hardware providing the collision indication in station X would have no way of detecting it, and, therefore, could not trigger the retransmission of the frame. The conclusion is that the frame duration has to be longer than τ.

As for the frame check sequence (FCS) field, it contains a four-byte cyclic redundancy check (CRC) value. Its value is computed as a function of the content of the source address, destination address, length, data and pad fields. The algorithm employed is based on the division of those fields by a known polynomial $G(x)$, the remainder determining the CRC. At the receiving side, $G(x)$ is employed to recompute the CRC value, which is compared against the FCS received value. If the two values coincide, the frame is declared valid, otherwise, it is not.

Let us complete this description by introducing and commenting on a further possibility that the 802.3 standard provides. We are referring to the MAC option of virtual LAN (VLAN) tagging, that reflects a frame format slightly different from the one just commented on. When this option is utilized, there is an additional four-byte VLAN header, called the QTag prefix, inserted between the source address and length/type fields, as Fig. 3.10 shows. The first two bytes of the prefix bear a specific reserved value, 0x81-00, informing the MAC that reads the frame that this is a special VLAN frame. The following two bytes allow specification of different transmission priorities for outgoing frames, ranging from 0 (the lowest) to 7 (the highest), as well as identification of the VLAN over which the frame has to be sent, a concept that we will meet in Chapter 5. For the time being, we simply enumerate the most interesting capabilities offered by VLAN tagging: through the priority mechanism, it is possible to assign higher transmission priority to frames with real-time constraints; moreover, stations can be logically grouped in VLANs, and will be able to communicate across multiple LANs as though they were on a single LAN.

To conclude, two major characteristics of the 802.3 MAC sublayer have to be emphasized: to provide a *connectionless*, *unacknowledged*, hence unreliable, type of service. As a matter of fact, in both the half-duplex and full-duplex modes that the IEEE 802.3 MAC supports we did not encounter the burdensome opening and closing connection phases, nor did we mention any acknowledgment for the transmitted frame. Once the frame reaches its local destination without incurring a collision, the MAC task ends: if the CRC procedure indicates that the frame was correctly received, no acknowledgment is sent back to the source; if an error is detected, the 802.3 MAC does not ask for a retransmission, it has no means to do it! The frame is simply discarded and its loss will be directly perceived by the network layer above.

Honestly, however, the standards imposed on the underlying physical medium the requirement to guarantee a low bit error ratio (BER), defined as the ratio of the number of bits received in error to the total number of bits received. Typical BER reference values lie in the 10^{-7}–10^{-10} range, so it is definitely not that frequent for the receiving NIC to run into an invalid frame, whose bits were altered during transmission.

3.2.3 Physical layer

3.2.3.1 Ethernet

The 802.3 standard refers to a variety of physical transmission media that can be employed at the LAN physical layer: coaxial cable (an obsolete solution), twisted pair and optical fiber (currently the two most widespread media), typically adopted in different contexts and when different needs arise, as Section 3.3 will show. Because the evolution of the Ethernet from its first versions up to the newest standard release is indissolubly tied with the physical layer of these networks, we go through its different specifications next.

After the original Ethernet networks operating at 1 Mbit/s, the next family of solutions exhibited a transmission rate of 10 Mbits/s: among the corresponding IEEE specifications, the one still worth being mentioned goes under the name 10BASE-T, a term where "10" recalls the supported bit rate, "BASE" indicates baseband transmission and "T" resembles the physical media employed, the unshielded twisted pair (UTP) cable.

The 10BASE-T specifications mandate the use of Manchester encoding at the physical layer, as did previous solutions. Fig. 3.11 visually describes it, highlighting the presence of signal level transitions at the middle of each symbol time: this feature helps the recovery of signal synchronization at the receiver, and also the detection of collisions. In 10BASE-T the original bus and point-to-point topologies are, for the first time, put side by side with the star topology, the reference solution in current Ethernet networks: the

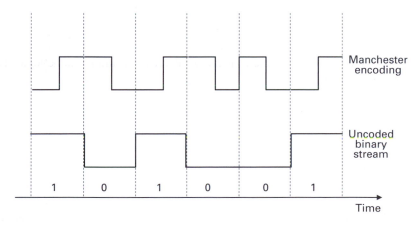

Fig. 3.11   Manchester encoding

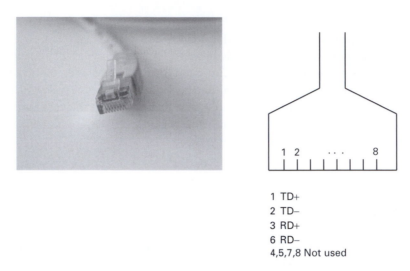

1 TD+
2 TD−
3 RD+
6 RD−
4,5,7,8 Not used

Fig. 3.12   The RJ-45 connector and plug, with the electrical scheme

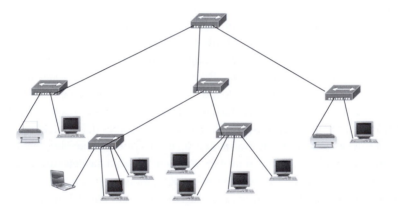

Fig. 3.13   A 10BASE-T Ethernet LAN employing several hubs

generic station connects to the central hub via two pairs of Category 3 (or better) UTP cables, terminated at both ends with an eight-pin RJ-45 connector, shown in Fig. 3.12. What does Category 3 stand for? We anticipate that the category indicates the quality of the cabling system, and takes on values ranging from 1, the lowest, to 6, currently representing the best quality (although Category 7 is already on the way); the last section in the chapter will elaborate in a systematic manner on the features that cabling systems exhibit, and on the guidelines to follow when an Ethernet LAN has to be installed and maintained.

There is a limit to the maximum distance that separates the hub from the station: in 10BASE-T it is equal to 100 m. Moreover, the diameter of the overall collision domain that 10BASE-T LAN exhibits cannot exceed 500 m: the last constraint applies to the topology that Fig. 3.13 exemplifies, where more stations are clustered under different hubs, in turn organized into a tree. An additional configuration criterion, known as the 5-4-3 rule, appears in 10BASE-T (as it did in the first 802.3 specifications): any path through a collision domain can, at most, traverse five segments, cross, at most, four repeaters (recall that hubs are multiport repeaters) and go through, at most, three segments that contain stations. Actually, all paths in 10BASE-T Ethernet networks with star topologies contain only two stations, as the previous figure clearly indicates.

As the processing power of the computer connected to the LAN gradually increased, the 10 Mbit/s bit rate became a bottleneck, compelling the 802.3 working group to deliver new, better performing standards, whose distinctive feature is to support a transmission rate of 100 Mbit/s. Here begins the adventure of fast Ethernet, whose story we tell next.

3.2.3.2 Fast Ethernet

Fast Ethernet, as it is commonly known, is technically described in the IEEE 802.3u addendum to the original 802.3 documents. It contains the 100BASE-TX, 100BASE-FX, 100BASE-T4 and 100BASE-T2 specifications, which all share the common feature of supporting the transmission rate of 100 Mbit/s.

The frame format in fast Ethernet does not undergo any significant change; in the half-duplex transmission mode, the channel access technique is still the 1-persistent CSMA/CD; however, the bit time is reduced by a factor of ten, not a trivial issue to consider when assessing the maximum diameter of the collision domain. As a matter of fact, detecting a collision when minimum-size frames are transmitted in a fast Ethernet LAN should take less than 5.12 µs: hence, it is not possible to stretch the LAN very far to guarantee a maximum propagation delay of this order of magnitude. Depending on the type of cable or fiber, and also on the hub characteristics, the diameter of a fast Ethernet collision domain can span from 200 to 320 m, but it cannot exceed these limits.

As for the different ways of implementing the fast Ethernet, 100BASE-TX is probably the most popular solution: it mandates the use of Category 5 UTP cables and supports a single cable length of up to 100 m. One major modification at the physical layer is that Manchester encoding is abandoned, mainly because it is too demanding in terms of bandwidth requirements, in favor of the 4B/5B encoding scheme, which maps four bits into a sequence of five bits. Out of the 32 available codewords, 16 correspond to data encoding, whereas some of the remaining 16 are reserved for special control purposes. As an example, the codeword 11111 is continuously sent over the media when no frames have to be transmitted, to guarantee continuous synchronization between the transmitting and receiving NICs: indeed a completely different approach from that used for the Ethernet LANs operating at 10 Mbits/s. Moreover, the two consecutive codewords 11000 and 10001 replace the preamble of the original frame format; the codewords 01101 and 00111 are appended at the end of each frame, after the FCS field, and act as a delimiter.

In 100BASE-TX, both half-duplex and full-duplex transmissions are supported.

The 100BASE-FX specification shares the same 4B/5B encoding scheme as 100BASE-TX, although it relies upon the usage of two multimode optical fibers, one for data transmission and the other for data reception. As the speed at which light propagates is higher than the speed at which an electrical signal travels across a copper cable, the propagation delay decreases and the length of a single fiber segment can be 412 m for half-duplex communications, i.e., when the CSMA algorithm is employed. When 100BASE-FX is used in full-duplex mode, the fiber segment can be 2000 m long.

The 100BASE-T4 and 100BASE-T2 specifications are additional specifications, both introduced to allow fast Ethernet to operate on existing Category 3 cabling systems.

The 100BASE-T4 specification uses all four pairs of the UTP cable that reach the user outlet, but it only supports half-duplex transmission.

The 100BASE-T2 specifications were developed to offer a better alternative than 100BASE-T4: the intent was to guarantee total backward compatibility with existing 10 Mbit/s solutions on UTP, therefore providing the 100 Mbit/s data rate on only two wire pairs. The corresponding products never made it to the market: the standardization process was too long, the engineering solutions too complex, and 100BASE-TX had all the time to spread and become the reference solution. Nevertheless, the 100BASE-T2 efforts laid out the basis for the next standard, the gigabit Ethernet, potentially taking the 1 Gbit/s data rate right to the user's desk.

Before stepping into this new realm, it is, however, worth dedicating a few words to the present-day solution of replacing the hub employed in the star topology of fast Ethernet LANs with a better performing device, commonly termed a *layer-2* switch.

What is a switch? Network fans can correctly interpret this device as a high performance bridge, i.e., a device that is more intelligent than the plain repeater, encompassing functionalities that stretch out of the physical layer and deeply involve the MAC. We defer the accurate treatment of bridges to Chapter 5, and confine our attention to a high-level understanding of the switch features and advantages.

When in the simple, reference star topology of Fig. 3.6, the switch replaces the hub, and the notion of collision domain disappears. Why is it so? Because all NICs are configured for full-duplex transmission, so that every station in the star can transmit and receive simultaneously, and so does every port of the switch; more importantly, each port of the switch is equipped with enough memory to buffer the incoming frame: following a suitable time schedule, when the frame arrives at the switch it is forwarded only toward the intended destination port via a high capacity backplane bus and then transmitted to the destination without causing any collision.

Equivalently, in a switched LAN, stations can send a frame at any time, provided the interframe gap is respected; from a logical standpoint, there are as many collision domains as the number of the switch ports, and there is only one station in each domain.

Switches can significantly boost performance: e.g., a 12 port switch for fast Ethernet can, in principle, reach an aggregate data transfer rate of 2.4 Gbit/s, although the actual data throughput will depend on its internal architecture. There should, however, be no doubt that for a switch the flow control mechanism outlined when the full-duplex transmission was first introduced is of great importance, as it allows the switch and the station to discipline data flow at layer-2, easing the task of upper layer protocols.

3.2.3.3   Gigabit Ethernet

New network applications are bandwidth eager, so it makes no surprise that the next step was to push the data rate limit even further: the process that led to the 802.3z gigabit

Ethernet standard started in 1995, with the last specification being published by IEEE in 1999.

The corresponding solutions, running at the impressive pace of 1 Gbit/s, retain backward compatibility with their previous 10 Mbit/s and 100 Mbit/s forerunners only for the full-duplex transmission mode, the one of most interest for gigabit Ethernet.

As for the half-duplex transmission mode, employing the 1-persistent CSMA/CD technique at this high data rate is no joke: if no additional countermeasures were adopted, the requirement of detecting all collisions, even those involving the minimum size frames, would force the collision domain to shrink by an additional factor of ten. To overcome this unbearable limitation, a frame-bursting technique has been introduced, consisting of several actions that the gigabit Ethernet NIC operating in half-duplex mode has to perform. To detect all collisions, extension bits are added at the end of small frames to bring them to the "new" minimum size of 512 bytes; in addition, to boost performance, once a station has gained access to the channel after the successful outcome of the carrier sense process, it is allowed to send more frames in a row, up to a maximum of 8192 bytes.

Honestly though, the half-duplex gigabit Ethernet is not of great interest; the full-duplex mode is the one that current products implement, together with the optional flow control mechanism: by transmitting a special frame, termed pause, it is possible to instruct the source not to send any more frames for a given amount of time.

The gigabit Ethernet specifications are: 1000BASE-T for UTP copper cable, Category 5 or better; 1000BASE-CX for shielded twisted pair (STP) copper cable, 1000BASE-SX for multimode optical fibers and 1000BASE-LX for both single-mode and multimode fibers.

All gigabit Ethernet versions support half- and full-duplex transmission, encapsulate each transmitted frame with special start of stream and end of stream delimiters, and maintain loop timing, continuously transmitting special idle symbols during interframe gaps, a solution already encountered in fast Ethernet.

The 1000BASE-T physical layer has the peculiar feature of employing forward error correction (FEC) coding to facilitate signal recovery in the presence of noise and other types of interference. When full-duplex transmission is considered, all four wire pairs are employed. The physical layer of the 1000BASE-X family of specifications employs a different coding scheme, based on ANSI fiber channel 8B/10B, which maps each data byte (eight bits) into a ten bit codeword. As for the limitations on the fiber length, in the full-duplex mode where the CSMA/CD mechanism is disabled, they are very relaxed when compared with the tight requirements of the original Ethernet: e.g., in 1000BASE-LX, strands of single mode optical fibers can be as long as 10 km, definitely an impressive value when compared with previous implementations.

3.2.3.4 The auto-negotiation option

Owing to the great variety of Ethernet interconnecting solutions we just described, the problem of automatically configuring the NICs at the two ends of a link, so that they can communicate under the same operational mode, has become of paramount importance.

The link start-up and initialization procedure known as auto-negotiation solves it, as it allows an NIC to advertise its Ethernet version, as well as to accept or reject the operational mode of the NIC at the other end, and to configure itself for the highest level mode that both NICs support.

With the goal of understanding its working principles, let us consider the auto-negotiation function in UTP-based NICs. In the initialization phase, the NIC sends bursts of fast link pulses (FLPs), wherein the clock sequence alternates with data bits: these declare the operational modes supported by the NIC and also bear additional information regarding the handshake mechanism that the auto-negotiation procedure employs. If the NIC at the other end of the link is compatible, but does not possess the auto-negotiation capability, a parallel detection function still allows the auto-negotiation to be recognized. On the contrary, a NIC that fails to respond to FLP bursts is interpreted as the lowest possible entry, i.e., a 10Base-T half-duplex NIC.

The advantages of auto-negotiation are tangible: the development of low-cost, multi-speed NICs, supporting, e.g., 10BASE-T and 100BASE-TX, in both half- and full-duplex operation. At the very least this translates into the advantage of softer network upgrades, as stations equipped with heterogeneous NICs can coexist within the same 802.3 LAN until the network administrator decides to replace older hardware with newer products.

3.3   Twisted-pair cabling standards

One of the main issues behind the proper implementation and maintenance of Ethernet networks is how to create the physical infrastructure that links user devices and supports data exchanges. A proper design and installation of such infrastructure, which involves aspects belonging to layer-1 of the OSI model, requires the solution of several distinct problems: they range from the type of cables employed, to the signal quality they guarantee, to the type of connectors and plugs employed, to the most adequate physical topology that the LAN should exhibit.

Local area network cabling has been an important topic since the beginning of the 1980s, when PC-based networks began to replace the terminal-server based ones. The first cabling systems appeared in those years, although they were proprietary solutions, tightly related to a given manufacturer that provided the whole network, from the plug all the way up to the application. At the end of that decade it became clear that a cabling system reference was necessary, the term "reference" indicating a set of rules and recommendations to regulate a given subject, with the aim of providing clear and fixed points despite technological advances. In 1991, the American standard organization EIA/TIA ratified the document *568 Commercial Building Telecommunication Standard* [35], which quite soon became the first world-wide recognized standard for the cabling of commercial buildings. This standard defines a multi-vendor and multi-product cabling system suitable to be installed during the construction or refurbishment of buildings and offices. It contains specifications for the cabling of a building or a set of buildings within a limited area (campus). It provides detailed recommendations regarding the transmission media, the topologies, maximum distances, connectors, backbones, installation, cable testing and certification.

The EIA/TIA identifies the quality of the copper wires and of the whole cabling system through categories, represented by an integer that currently ranges from 1 to 6 (from low to high quality), with an additional Category 5e, where the "e" stands for "enhanced." At the international level, however, ISO/IEC 11801 [36] is the cabling standard. It classifies cables, components and communication links over copper wires with a letter: Class A (low quality), B, C, D, E and F (high quality). This standard reports specifications for the electrical properties, testing procedures and recommendations for electromagnetic compatibility (EMC), including shielding and grounding procedures. On the European side, EN 50173 [37] is the standard for cabling, similar to ANSI EIA/TIA 568, yet coming from ISO/IEC 11801. The EN 50173 standard mainly refers to European standards, it does not have the equivalent of EIA/TIA Category 4 cables and it reports detailed procedures for link testing and performance.

The ANSI document EIA/TIA 568 is the standard that this section will mainly present. The basic topology it introduces is a hierarchical star, encompassing several elements of the cabling system. Figure 3.14 is an example of this topology for a three-building campus, and represents a good starting point to introduce the fundamental blocks that constitute it. The main cross-connect (MC) is the campus star center, distributing data among the different buildings. It is the center of the so-called inter-building backbone. In each building the intermediate cross-connect (IC) is the local star center, the mid-point of the intra-building backbone connecting all floors. On each floor the telecommunication closet (TC) is the floor star center, connected to the IC through the intra-building backbone: all the telecommunication outlets (TO) that equip the user workplaces are connected to the TC via point-to-point cables.

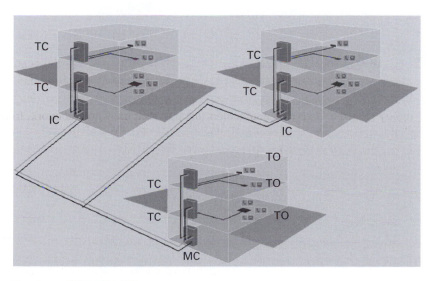

Fig. 3.14   Topology of EIA/TIA 568

As regards horizontal distribution, the maximum distance between each TO and the TC is 90 m. In addition, patch cords up to 3 m long can be employed in each office, to connect the station NIC to the TO, as the picture in Fig. 3.15 illustrates; this picture also indicates that two outlets per workplace are usually installed and that the reference connector is the RJ45 connector introduced in previous section. Patch cords with the same limitation in length are also used in the floor TC: here, each cable coming from a TO is terminated in a patch panel, and a patch cord allows it to reach the port of the active network device that the station needs to be connected. Figure 3.16 introduces the main elements behind this description, whereas the picture in Fig. 3.17 shows a typical, fairly crowded situation that can be encountered in a TC.

There are many advantages related to the adoption of the EIA/TIA 568 topology, in particular as regards network management. A hierarchical star significantly helps network administrators to handle, for instance, the insertion and removal of stations by properly configuring the TCs. In addition, fault detection is much easier, as an out-of-service link only impacts on the user or users connected to that link; moreover, locating faults and shortening the out-of-service times becomes simpler.

Fig. 3.15   Detail of a workplace with the patchcord between the NIC and the telecommunication outlet TO

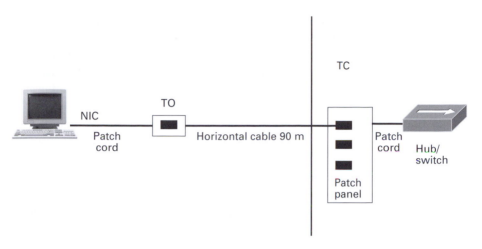

Fig. 3.16   Main elements of the horizontal cable distribution

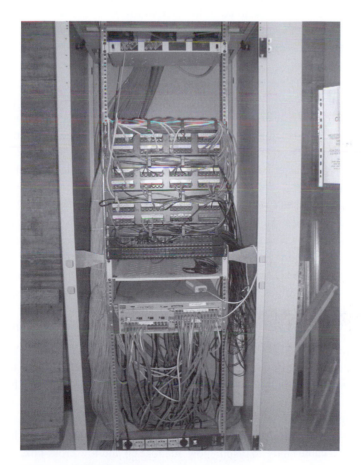

Fig. 3.17   A telecommunication closet

So far for the topology. Now another significant topic awaits us: the assessment of the quality of the cabling system, which ultimately determines the achievable transmission rate, and hence the type of Ethernet network that can be implemented on top of it. As a matter of fact, the higher the bit rate the system is asked to support, the more stringent the requirement on the cabling system. To understand it, it is sufficient to focus on a single cable: to carry a signal whose data rate is 100 Gbit/s will definitely require more layer-1 bandwidth than that needed by a 10 Mbit/s signal.

We will see that several parameters have to be measured to certify the cable quality and the quality of the cabling system as a whole. The reference standard for testing and certifying UTP cables is the ANSI EIA/TIA TSB67 [38] technical services bulletin, that can be used for STP cables too. This document defines two main reference settings for cable testing: the basic link and the channel. The first term indicates the permanent path between the TO and the TC, excluding patch cords; the second term points to the entire communication path between the station and the hub or switch within the TC, including the horizontal distribution cable, the patch panel and all the patch cords.

Generally speaking, a good transmission medium should have low resistance, low capacity and low inductance or, to put it in other words, it should be neither dispersive nor dissipative: most of the transmitted power should get to the receiver and the received signal should be neither distorted nor reflected, nor spoiled by interference that other signals might introduce.

To guarantee this, UTP cables have to meet specific criteria in the desired layer-1 bandwidth in terms of attenuation, cross-talk and impedance, that the following paragraphs will gradually define. These electrical properties depend on the geometric characteristics of the examined cable, such as the diameter and the number of the wires in the cable, their distances from each other, and the presence of metallic shields. They also depend on the wire material, characterized by a dielectric constant, a magnetic permeability index and a conductance.

As regards UTP (and STP) cables, the fundamental geometric indication that needs to be specified is the diameter of the copper wires they are made of: its measure allows one to determine the wire resistance when the resistivity is known and also to understand how easily the cable can be bent, as thinner wires are more flexible. The diameter size is traditionally measured in AWG (American wire gage), the AWG being a geometric regression scale that employs 39 values spanning the 000–36 range, 000-AWG corresponding to a 0.460 inch diameter, 36-AWG to a 0.005 inch diameter. Note that thicker wires have smaller AWG values and that the increment of one gage corresponds to a ratio between diameters of $(0.460/0.005)^{\frac{1}{39}} = 92^{\frac{1}{39}} = 1.1229322$. So, 26-AWG and 24-AWG correspond to a wire diameter of 0.0159 and 0.0201 inches, respectively (0.5106 and 0.4049 mm in the more appealing international metric system). The 24-AWG size is the usual wire size of Ethernet cables (Category 3, 4, 5 and 5e), while 26-AWG is the typical diameter of the wires used in patch cords; thinner as they need to be more easily bent.

Besides the physical property of good flexibility, to allow the transmission of high data rate signals, cables have to display low attenuation and low cross-talk.

The attenuation or insertion loss indicates how much of the signal power is lost along the line. It is defined as the ratio between the transmitted power P_{in} and the received power P_{out}, and it is traditionally given in dB:

$$\text{Insertion\_Loss[dB]} = 10 \log_{10} \frac{P_{in}}{P_{out}}.$$

The insertion loss increases as a linear function of the frequency.

As for the cross-talk, this phenomenon consists of the undesired presence on a pair of wires of a signal coming from another close pair, that interferes with the reference signal. The cross-talk in dB is expressed as:

$$\text{Cross\_talk[dB]} = 20 \log_{10} \frac{v}{v_{ext}},$$

where v is the information signal on a pair of wires, while v_{ext} is the fraction of the signal that spills over a nearby pair. This cross-talk measure has to take on high values: IEEE 802.3 physical layer specifications set minimum values for it. In 10BASE-T, minimum cross-talk values were 30.5 dB at 5 MHz and 26 dB at 10 MHz for four pairs of cables, typically adopted in horizontal distribution, and 35 dB at 5 MHz and 30.5 dB at 10 MHz for 25 pairs of cables, which can be employed in intra-building backbones. When v is measured at the transmitter side and v_{ext} at the receiver side of the close pair of wires, the cross-talk is usually referred to as pair-to-pair near-end cross-talk (NEXT). To quantify the pair-to-pair far-end cross-talk (FEXT) is also important: this additional measure takes into account the interference between adjacent pairs when the interference on the outgoing signal is caused by the incoming signal on a close pair. In this case v_{ext} is measured at the transmitter side of the close pair. Both NEXT and FEXT are graphically illustrated in Fig. 3.18.

There are two main methods to measure the cross-talk. The first is called pair-to-pair: the maximum interference generated by a single active pair is measured and taken into account; if more pairs are active, performance will be worse than that predicted. The second is called power sum (PS): measurements are carried out when all pairs in the cable are active; for cables with more than four pairs, this is considered the only reliable way to measure cross-talk.

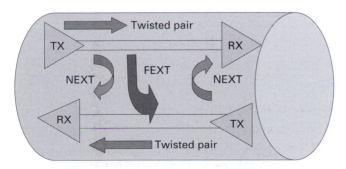

Fig. 3.18   NEXT and FEXT

Absolute values of attenuation and cross-talk are usually not as interesting as their combination. If external noise is supposed negligible, it is their combination that determines the signal-to-noise ratio at the receiver input, i.e., the actual quality of the received signal. The attenuation to cross-talk ratio (ACR) is the ratio between the insertion loss and NEXT. In dB, ACR(dB) = Insertion_Loss[dB] − NEXT[dB]: signal quality is significantly degraded when this difference takes on positive values that are relatively high. As an example, for a Category 6 Channel, at 1 MHz, the maximum insertion loss is 2.1 dB and the minimum NEXT is 65 dB (ACR = −62.9 dB); at 250 MHz the maximum insertion loss is 35.9 dB and the minimum NEXT is 33.1 (ACR = +2.8 dB).

The attenuation to cross-talk ratio far end (ACRF), also called equal level far-end cross-talk (ELFEXT), is, on the other hand, the ratio between the insertion loss and the FEXT. Therefore, as with ACR, it is important to consider the difference Insertion_Loss[dB] − FEXT[dB], which turns out to be independent of the cable length.

Another significant parameter to test for determining the cable quality is its impedance, which is related to its physical properties. A wire or cable may present irregularities in its geometry that cause the electrical characteristics to change. When an electrical signal meets them, it experiences reflections that in turn make the signal lose energy. The loss due to reflections is called return loss. An additional, severe consequence of reflections is the rise of jitter, that can cause errors at the receiver. The reference impedance values for UTP cables is 100 Ω.

Propagation delay is a physical property of a transmission medium: in Chapter 1 we stated that it depends on the propagation speed of a signal in the medium and on the distance the signal travels. In a multi-pair cable it is important to measure this delay for each pair, as there might be some significant skewness between pairs: as some Ethernet solutions such as fast Ethernet(100BASE-T4) and gigabit Ethernet(1000BASE-T) may use all four pairs for transmissions, the excessive delay skew may lead to transmission errors.

In conclusion, to assess the quality of a cabling system, it is necessary to certify it with suitable measurement instruments. These instruments are properly connected to the cabling system to test some or all of the above parameters. The results of the corresponding measurements determine the cabling Category or its Class.

Category 5 and Category 5e (enhanced) cables are certified to operate in a bandwidth up to 100 MHz. Some cables might also be marked as "Cat.5e XXX MHz tested," where XXX stands for a value of tested bandwidth higher than 100 MHz (e.g., 350 MHz).

The EIA/TIA 568-B.2 standard document released in 2001 describes Category 5e [39]. It contains more requirements than EIA/TIA 568 (referring to Category 5), and considers all components of the cabling system (cables and also outlets, patch cords, patchpanels, . . .). It also specifies several new testing terms that are fundamental when all four pairs of a cable are simultaneously employed for transmission in one direction. The list of terms reported in TIA/TSB-67 had then to be extended for dealing with all the four pairs, and another document, TSB-95 [40], has been approved for reporting the new measurements to carry on, such as the previously mentioned FEXT, Power Sum FEXT and ACRF.

The Category 6 standard was published in 2002. The corresponding document requires that each cabling component be Category 6: the channel bandwidth is raised to 250 MHz and all components are tested at 250 MHz. As regards the horizontal cabling, Category 6 maintains what Category 5 recommended: four twisted pairs per cable, unshielded (UTP), shielded (STP) or foiled (FTP). The maximum distance between the TC and the TO is 100 m. Electrical parameters to be tested for a Category 6 certification are reported in the EIA/TIA-568-B.2-1 document [41]. Technical bulletins TIA TSB-155 [42] and ISO/IEC 24750 [43] report technical procedures for certifying a Category 6 cabling system, corresponding to ISO/IEC Class E cabling, that is able to support not only 100BASE-T but also 1000BASE-T.

Finally, the standard TIA/EIA-568-B.2-Addendum 10 [44] reports electrical specifications for Category 6a ("a" stands for "augmented"), imposing more restrictive bounds for the electrical parameters, with the goal of providing a bandwidth up to 500 MHz. The principal aim is to support 10GBASE-T communications.

On the ISO/IEC side, Class F specifications have been published in 2002 within the second edition of the 11801 document standard: they recommend an entirely shielded cable, with global and per-pair shields. The ISO/IEC Class F should correspond to TIA Category 7, even if at the time of this writing TIA has not approved a document describing its requirements yet. The goal of Class F is to support communications in a bandwidth up to 600 MHz. These specifications are currently evolving toward Class Fa (augmented), which extends the bandwidth until 1000 MHz for supporting, for instance, not only high speed data networking but also applications that are typical of a completely different setting, namely, television broadcasting.

Table 3.1 highlights the correspondence between the cabling system Category or Class and the reference standard documents published by EIA/TIA and ISO/IEC.

3.4 Practice: address resolution protocol

We now abandon the wordy description of cables to return to the practical issue of addressing in LANs. As we mentioned before, the NIC is the device that allows the host to be connected to the network, according to the physical and data-link layer specifications implemented in the NIC itself. The addressing scheme in the MAC sublayer mandates that each NIC be assigned a low-level address, which we equivalently termed hardware address, MAC address or physical address.

In Chapter 2, however, we spent quite a few pages in convincing the reader that network administrators assign IP addresses to hosts (now we could say NICs), according to network layer numbering requirements.

The two addressing schemes are applied at separate layers and are independent of each other. How do we match them, as we undoubtedly need to do so? A specific mechanism has to be adopted to establish the correspondence between MAC and IP addresses assigned to a NIC.

Table 3.1. Main cabling standards

TIA cabling standards	
Category 5e	ANSI/TIA/EIA-568-B.2, Commercial Building Telecommunications Standard Part2, 2001
Category 6	ANSI/TIA/EIA-568-B.2-1, Commercial Building Telecommunications Standard Part2: Add. 1, 2002
Category 6a	ANSI/TIA/EIA-568-B.2-10 (draft), Commercial Building Telecommunications Standard Part2: Addendum 10
ISO cabling standards	
Class D	ISO/IEC 11801, 2nd Ed., Information Technology Generic Cabling for Customer Premises, 2002
Class E	ISO/IEC 11801, 2nd Ed., Information Technology Generic Cabling for Customer Premises, 2002
Class Ea	Amendment 1 to ISO/IEC 11801, 2nd Ed., IT Generic Cabling for Customer Premises, pending
Class F	ISO/IEC 11801, 2nd Ed., Information Technology Generic Cabling for Customer Premises, 2002
Class Fa	Amendment 1 to ISO/IEC 11801, 2nd Ed., IT Generic Cabling for Customer Premises, pending

Let us see how this is automatically triggered in current networks, focusing on the case of one host willing to send an IP datagram to a single destination. In this circumstance, the sender knows both its own MAC address and IP address. It also knows the receiver's IP address: for the case of direct delivery discussed in Section 2.4.4, it is the destination address specified in the IP packet header; for the case of indirect delivery it is the IP address of the next hop, e.g., the default gateway.

The problem is to find the MAC address of the receiver. This task is performed by the *address resolution protocol* (ARP). The most common ARP implementation is the one defined in RFC 826 [45] for the case of LANs based on the Ethernet/IEEE 802.3 standard, which we can generically frame as "multi-access networks supporting broadcast transmissions."

With reference to the example in Fig. 3.19, the protocol works as follows:

(i) The sender transmits a MAC broadcast frame carrying an *ARP request* where the target IP address (137.204.57.26) is specified; the sender's IP (137.204.57.174) and MAC (00:11:43:D1:18:9A) addresses are also specified;

(ii) Every station connected to the same LAN receives the request, but ignores it unless the target IP address matches its own address; in this case, the target station (137.204.57.26) sends a unicast *ARP reply* directly to the host that raised the request, specifying its physical address (00:80:9F:6F:C1:DC);

(iii) The sender now knows the MAC address of the target and is able to transmit frames to it; it additionally adds an entry to a specific ARP table stored in its memory with

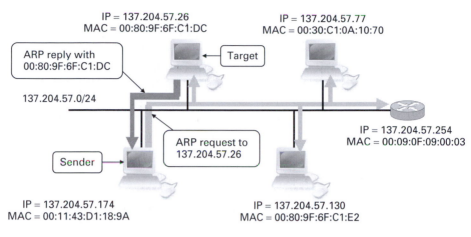

Fig. 3.19   Address resolution protocol

the target's IP and physical address couple (137.204.57.26, 00:80:9F:6F:C1:DC), so it will not need to trigger the ARP request-reply procedure again, in case it has further data to send to this target;

(iv) Similarly, the receiver adds the sender's IP and physical address couple (137.204.57.174, 00:11:43:D1:18:9A) to its own ARP table: this eliminates the transmission of a further ARP request in the very likely case that the receiver has some data to transmit back to the sender (e.g., a reply required by higher-layer protocols).

   At both sender and receiver side, each ARP transaction generates a new entry in the ARP table. Every time a new packet has to be sent, the sender checks whether the target's physical address is stored in its ARP table; if not, the ARP procedure is initiated. However, the correspondence between IP and MAC address might be subject to changes in time, for instance because a station has been assigned a different IP address while keeping the same hardware, or because its NIC has been replaced owing to some failure or to an upgrade. As a consequence, since the information kept in the ARP table may become outdated, each entry has a limited lifetime and when this expires the entry is deleted, forcing the sender to initiate new ARP transactions to keep the table updated.

   Major operating systems provide commands to show and manipulate the local ARP table. On a Linux box, this can be done via the `arp` command, while the `arping` command allows generation of custom ARP requests.

   As an example, invoking the `arp` command on the sender in Fig. 3.19 before the ARP transaction takes place returns the following ARP table:

```
[root@linuxbox ~]# arp -n
Address            HWtype   HWaddress           Flags Mask     Iface
137.204.57.130     ether    00:80:9F:6F:C1:E2   C              eth0
137.204.57.254     ether    00:09:0F:09:00:03   C              eth0
137.204.57.77      ether    00:30:C1:0A:10:70   C              eth0
```

In each of its entries there appears the indication of the IP address, of the type of underlying network (an Ethernet LAN in this example), of the corresponding MAC address, an additional piece of information we intentionally skip (flags mask) and the indication of the network interface (labeled eth0 in the example).

The ARP request shown in Fig. 3.19 can be forced by the command:

```
[root@linuxbox ~]# arping -c 1 -I eth0 137.204.57.26
ARPING 137.204.57.26 from 137.204.57.174 eth0
Unicast reply from 137.204.57.26 [00:80:9F:6F:C1:DC]  3.343ms
Sent 1 probes (1 broadcast(s))
Received 1 response(s)
```

The next inspection of the sender's ARP table highlights the following update:

```
[root@linuxbox ~]# arp -n
Address         HWtype  HWaddress          Flags Mask    Iface
137.204.57.130  ether   00:80:9F:6F:C1:E2  C             eth0
137.204.57.254  ether   00:09:0F:09:00:03  C             eth0
137.204.57.77   ether   00:30:C1:0A:10:70  C             eth0
137.204.57.26   ether   00:80:9F:6F:C1:DC  C             eth0
```

The ARP behavior illustrated above refers to the case of a network that supports broadcast transmission. However, when the underlying multi-access network does not allow broadcast transmissions, a modified version of the protocol must be adopted, which translates into the presence of one or more servers responsible for maintaining an updated ARP table for the whole network. The ARP requests are then sent to those servers that reply with the IP-to-physical address correspondence, given that this is available in their table. Of course, all clients are required to register their IP and physical addresses to such servers at boot time and to keep this information updated. A local ARP cache is also available in each client to improve efficiency.

An extension to the ARP mechanism in broadcast networks, described in RFC 925 [46] and often refered to as *proxy ARP*, allows a host to reply with its own MAC address to an ARP request actually directed toward a different station, as Fig. 3.20 illustrates. With the proxy ARP mechanism, it is possible to employ the same address resolution protocol on different network segments connected through a proxy device: it is the proxy that is in charge of replying with its own physical address to ARP requests coming from a sender connected to a different segment than the target. Therefore, IP packets addressed to the target are encapsulated by the sender in MAC frames directed to the proxy device. The proxy then takes care of forwarding the packets to the actual destination. This operation is completely transparent to the sender, which believes the target is on the same network segment. Proxy ARP is typically used to join different LAN segments or a LAN segment with a point-to-point link in order to make them appear as if they were a single LAN, as shown in Fig. 3.20. Unfortunately, it can also be misused by malicious hosts to intercept the traffic exchanged between two other hosts and perform packet sniffing and LAN-based man-in-the-middle attacks.

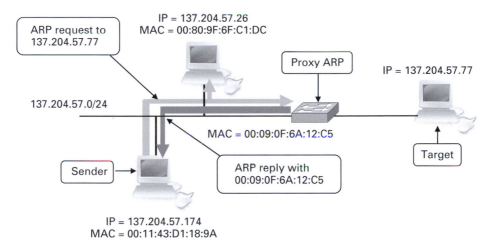

IP = 137.204.57.26
MAC = 00:80:9F:6F:C1:DC

ARP request to
137.204.57.77

Proxy ARP

IP = 137.204.57.77

137.204.57.0/24

MAC = 00:09:0F:6A:12:C5

Sender

Target

ARP reply with
00:09:0F:6A:12:C5

IP = 137.204.57.174
MAC = 00:11:43:D1:18:9A

Fig. 3.20   The use of proxy ARP to join a broadcast LAN and a point-to-point link

A Linux box can be configured to act as a proxy ARP device using the `arp` command. With reference to the example in Fig. 3.20 and assuming that `eth0` is the proxy ARP interface connected to the LAN, the required syntax is the following:

```
[root@proxy ~]# arp -i eth0 -s 137.204.57.77 00:09:0F:6A:12:C5 pub
```

To experiment with this, it is also necessary to know that to make things work, the packet forwarding function between different interfaces must be enabled on the Linux box. This is typically done by invoking the command:

```
[root@proxy ~]# sysctl -w net.ipv4.ip_forward=1
net.ipv4.ip_forward = 1
```

The address resolution problem discussed so far has assumed that every host knows its own IP and MAC addresses. However, this may not always be the case. As a matter of fact, the MAC address is known because it is permanently assigned to the network interface hardware, but a station may not be configured with a fixed IP address. To tackle this issue, the *reverse ARP* (RARP) protocol has been defined in RFC 903 [47], allowing a host to send a broadcast request asking for the IP address corresponding to its hardware address. One or more RARP servers have to be active on the same network and have to maintain a correspondence table, so as to reply to RARP requests with the IP address assigned to the client. In the past, RARP was used to configure diskless workstations at boot time, when hard disks were very expensive. Today, more advanced dynamic host configuration solutions are in place, such as the dynamic host configuration protocol (DHCP) defined in RFC 2131 [48].

3.5   Practice: NIC configuration

Each host – now we could say each NIC connected to a local IP network – willing to communicate via the TCP/IP protocol suite has to be initially configured with:

- A unique IP address belonging to the local network address space;
- A netmask corresponding to the local network size;
- The address of a *default gateway* on the local network, in case the NIC must be used to perform indirect deliveries of packets directed to remote IP networks.

This is the information required by the NIC configuration panel shown in Fig. 3.21 for a host running a Microsoft Windows™ operating system. Incidentally, this figure also shows the addresses of two domain name system (DNS) servers, which the host needs to know to solve the additional correspondence between symbolic host names and actual IP numerical addresses: being explicit, given we want to connect to a web server named ing.unimore.it, how does our machine learn its IP address? The first step to find it out is to send a query to the DNS servers our machine knows. This will in turn trigger a search in a distributed database, but we do not intend to dig into it any deeper.

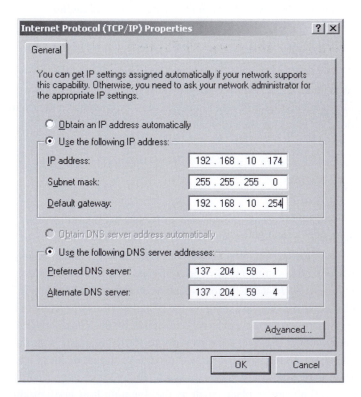

Fig. 3.21   NIC configuration under a Windows™ system

Under a Linux box, the NIC configuration information can be shown using the `ifconfig` command as follows:

```
[root@linuxbox ~]# ifconfig
eth0      Link encap:Ethernet  HWaddr 00:11:43:D1:18:9A
          inet addr:137.204.57.174  Bcast:137.204.57.255
          Mask:255.255.255.0
          UP BROADCAST RUNNING MULTICAST  MTU:1500  Metric:1
          RX packets:1091117 errors:0 dropped:0 overruns:0 frame:0
          TX packets:1260884 errors:0 dropped:0 overruns:0 carrier:0
          collisions:0 txqueuelen:1000
          RX bytes:108102787  TX bytes:1603504548

lo        Link encap:Local Loopback
          inet addr:127.0.0.1  Mask:255.0.0.0
          UP LOOPBACK RUNNING  MTU:16436  Metric:1
          RX packets:624023 errors:0 dropped:0 overruns:0 frame:0
          TX packets:624023 errors:0 dropped:0 overruns:0 carrier:0
          collisions:0 txqueuelen:0
          RX bytes:218515859  TX bytes:218515859
```

The command lists the NICs installed on the system, in our example an Ethernet card named `eth0` and the loopback virtual interface named `lo`. For each NIC, the IP configuration is also shown, including address, netmask and maximum transfer unit (MTU), as well as some traffic statistics. Additional information for the Ethernet interface includes the hardware address and the directed broadcast value.

The information about the default gateway that the NIC has been configured with can be retrieved by looking at the host routing table, shown by the `route` command:

```
[root@linuxbox ~]# route
Kernel IP routing table
Destination     Gateway         Genmask         Flags Metric Iface
137.204.57.0    *               255.255.255.0   U     0      eth0
default         137.204.57.254  0.0.0.0         UG    0      eth0
```

This command lists the kernel routing entries, including:

- The network-specific route related to the local IP network;
- The route for the default gateway.

In our example the local network route refers to prefix 137.204.57.0/24. Here the absence of a gateway address indicates that direct delivery must be performed for every packet addressed to a host on such network. For any other packet, the address of the default gateway to be used for indirect delivery is 137.204.57.254.

We note in passing that on a Linux box the information about the DNS servers' addresses is typically included in a configuration file:

```
[root@linuxbox ~]# cat /etc/resolv.conf
nameserver 137.204.78.3
nameserver 137.204.78.2
```

Since IP configuration is NIC-specific, multi-homed hosts – and routers as well – have to be properly configured on each interface. For instance, the configuration for a host with two NICs (eth0 and eth1) looks like:

```
[root@linuxbox ~]# ifconfig
eth0      Link encap:Ethernet  HWaddr 00:11:43:D1:18:9A
          inet addr:137.204.57.174  Bcast:137.204.57.255
          Mask:255.255.255.0
          UP BROADCAST RUNNING MULTICAST  MTU:1500  Metric:1
          RX packets:5278462 errors:0 dropped:0 overruns:0 frame:0
          TX packets:1401002 errors:0 dropped:0 overruns:0 carrier:0
          collisions:0 txqueuelen:1000
          RX bytes:2043018344  TX bytes:939090566

eth1      Link encap:Ethernet  HWaddr 00:0E:0C:5E:3F:1C
          inet addr:192.168.10.174  Bcast:192.168.10.255
          Mask:255.255.255.0
          UP BROADCAST RUNNING MULTICAST  MTU:1500  Metric:1
          RX packets:3571089 errors:0 dropped:0 overruns:0 frame:0
          TX packets:518981 errors:0 dropped:0 overruns:0 carrier:0
          collisions:0 txqueuelen:1000
          RX bytes:318990970  TX bytes:530986376

lo        Link encap:Local Loopback
          inet addr:127.0.0.1  Mask:255.0.0.0
          UP LOOPBACK RUNNING  MTU:16436  Metric:1
          RX packets:5256 errors:0 dropped:0 overruns:0 frame:0
          TX packets:5256 errors:0 dropped:0 overruns:0 carrier:0
          collisions:0 txqueuelen:0
          RX bytes:2437600  TX bytes:2437600
```

The corresponding host routing table showing two entries for direct delivery to the local networks and the single default route is:

```
[root@linuxbox ~]# route
Kernel IP routing table
Destination     Gateway         Genmask         Flags Metric Iface
137.204.57.0    *               255.255.255.0   U     0      eth0
192.168.10.0    *               255.255.255.0   U     0      eth1
default         137.204.57.254  0.0.0.0         UG    0      eth0
```

The ifconfig command can also be used to configure the parameters of an interface. For instance, the syntax to configure an Ethernet interface named eth2 with address 10.5.8.188 as part of network prefix 10.5.8.128/25 is

```
[root@linuxbox ~]# ifconfig eth2 10.5.8.188 netmask 255.255.255.128
                   broadcast 10.5.8.255
```

Note that the previous and the following commands must be typed on a single line. The route command is also used to set the default gateway, that in our example corresponds to the IP address 10.5.8.254:

```
[root@linuxbox ~]# route add default gw 10.5.8.254
```

Sometimes it is useful to have more than a single IP address assigned to the same NIC, for instance when multiple IP networks are used on the same physical segment and the host should be connected to some or all of them. Again, on a Linux box this can be achieved via the `ifconfig` command, configuring multiple *IP aliases* for the same NIC. For instance, the syntax to assign a main address and two aliases to `eth0` and to show the obtained configuration is

```
[root@linuxbox ~]# ifconfig eth0 137.204.57.174 netmask
                255.255.255.0 broadcast 137.204.57.255
[root@linuxbox ~]# ifconfig eth0:0 192.168.10.174 netmask
                255.255.255.0 broadcast 192.168.10.255
[root@linuxbox ~]# ifconfig eth0:1 10.0.0.74 netmask 255.255.255.128
                broadcast 10.0.0.127
[root@linuxbox ~]# ifconfig
eth0      Link encap:Ethernet  HWaddr 00:11:43:D1:18:9A
          inet addr:137.204.57.174  Bcast:137.204.57.255
          Mask:255.255.255.0
          UP BROADCAST RUNNING MULTICAST  MTU:1500  Metric:1
          RX packets:5278462 errors:0 dropped:0 overruns:0 frame:0
          TX packets:1401002 errors:0 dropped:0 overruns:0 carrier:0
          collisions:0 txqueuelen:1000
          RX bytes:2043018344  TX bytes:939090566

eth0:0    Link encap:Ethernet  HWaddr 00:11:43:D1:18:9A
          inet addr:192.168.10.174  Bcast:192.168.10.255
          Mask:255.255.255.0
          UP BROADCAST RUNNING MULTICAST  MTU:1500  Metric:1

eth0:1    Link encap:Ethernet  HWaddr 00:11:43:D1:18:9A
          inet addr:10.0.0.74  Bcast:10.0.0.127
          Mask:255.255.255.128
          UP BROADCAST RUNNING MULTICAST  MTU:1500  Metric:1
```

The commands shown above are used to configure IP parameters on a Linux box at runtime, but these configurations are not permanent and do not survive after a system reboot. To make them permanent, they have to be saved on some configuration files, which depend on the Linux distribution used. Similar command line tools on Microsoft Windows™ systems are `ipconfig`, `route` and `netsh interface ip`.

3.6 Practice: a campus network layout

Sketch the design of a campus network following the guidelines given by EIA/TIA-568-B. The campus network has three buildings of three floors each. Every floor has 16 offices and a laboratory. Each office has one workplace and in each laboratory there are 12 workstations. Every workplace has to be equipped with a workstation connected to the campus network and with a telephone.

Solution Figure 3.22 shows the design of the campus network, while Fig. 3.23 is a sketch of the design of each floor.

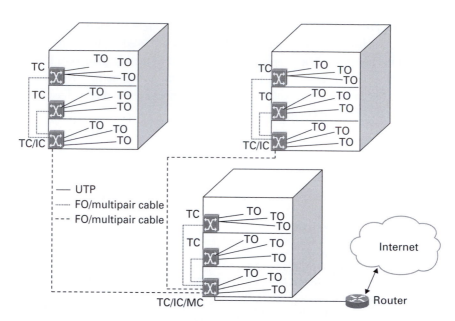

Fig. 3.22   Campus network design

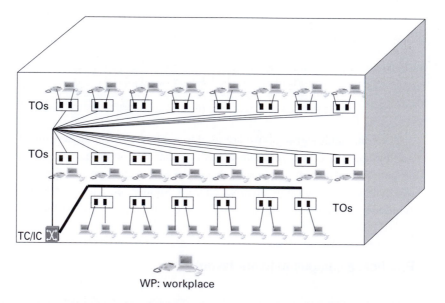

Fig. 3.23   Horizontal distribution on each floor

4   Wireless networks

The intent of this chapter is to complement the previous discussion on wired LANs, providing a thorough picture of the technologies and standards available today to build wireless network infrastructures. These aim to cover a wide range of areas successfully, spanning from personal to local and metropolitan coverage solutions.

The impressive pace of technological advances of the last decade has acquainted us with technologies that replace cable over very short distances, as testified by palm computers, cellular headsets and plenty of technological "toys" equipped with Bluetooth ports. Wireless LANs have an unprecedented popularity too, rapidly pervading private houses and enterprises; in metropolitan areas, municipalities are also getting into the game, supporting projects of wireless connectivity for communities that would otherwise be confined to low-speed Internet access. This all started with some *impressive* – at the time – 1 and 2 Mbit/s raw transmission rates that the first Wireless LAN standard guaranteed: current products achieve 54 Mbit/s of raw throughput, rates of 250 Mbit/s are on the way, and the specifications for wireless MAN head to even higher data rates, extended coverage and mobility support.

The logical organization of the chapter reflects the colloquial description above: the first section is devoted to wireless LANs and to the corresponding set of specifications, belonging to the IEEE 802.11 family of standards; the second section deals with the issue of fixed – and mobile – radio broadband access that wireless MANs have to provide. For the sake of completeness, the next sections rapidly look at wireless personal area networks (WPANs) and wireless mesh networks. The chapter is closed by some practical – and crucial – considerations on traffic monitoring and security issues in wireless networks.

4.1   Wireless LAN

4.1.1   The basket of 802.11 standards

The reference point for WLANs is represented by the IEEE documents, exactly as for LAN standards: 802.11 is the IEEE working group for wireless local area networks. The 802.11 documents can well be interpreted as the equivalent of 802.3 documents for wired LANs, and as such, they provide PHY and MAC layer specifications that manufacturers of WLAN equipment should adhere to.

The IEEE standards for WLAN have known quite an evolution, mainly regarding the physical layer specifications that the IEEE 802.11 working group ratified year after year. In chronological order, the working group produced the standard document 802.11, 1999 edition [49], providing PHY and MAC specifications for a WLAN operating at 1 and 2 Mbit/s on the 2.4 GHz unlicensed ISM frequency band: frequency hopping and direct sequence were the two spread spectrum techniques envisaged at the PHY layer, paired with a third alternative, infrared. The IEEE 802.11a document [50], for a WLAN operating at 54 Mbit/s on the 5 GHz band and adopting in its PHY layer the OFDM technique was available shortly after; it was paired by the IEEE 802.11b-1999 Supplement to 802.11-1999 [51], an extension of the original standard in the 2.4 GHz band, attaining bit rates up to 11 Mbit/s with complementary code keying (CCK) modulation; the IEEE 802.11g standard [52], encompassing a further higher speed extension in the 2.4 GHz band, which achieves 54 Mbit/s and relies on OFDM in its physical layer, was released in 2003. We will quickly go over OFDM in the section devoted to wireless MAN, as OFDM is the winning radio technology in this context.

The IEEE 802.11e and 802.11i documents are also worth mentioning [53, 54], because they target two pressing issues in WLANs: the need for quality of service and for security, respectively.

4.1.2   Physical layer evolution

A description of the main PHY layer features that the original 802.11, 802.11a, 802.11b and 802.11g WLAN standards encompass is provided next, with the intent to emphasize the endless quest for higher and higher data rates: not a trivial goal in the scarce radio spectrum.

To begin with, the first designated frequency band for WLAN: it is the 2.4 GHz band reserved for industrial, scientific and medical (ISM) applications, featuring, on the one hand, the appealing characteristic of being licence-free and, on the other hand, the drawback of being prone to interference from such devices as microwave ovens, Bluetooth devices and cordless telephones, which operate over the same frequencies.

The former 802.11 standard document dictates three alternative transmission scheme specifications: frequency hopping spread spectrum (FHSS), direct sequence spread spectrum (DSSS) and infrared (IR), supporting two data rates: 1 and 2 Mbit/s.

We spend a few words to explain what hides behind FHSS and DSSS. In frequency hopping systems, the radio transmitter hops in time from one frequency carrier to another, following a predefined sequence; accordingly, the intended receiver follows the same pattern to tune into the right subchannel and demodulate the desired signal.

In DS spread spectrum systems, the transmitted signal occupies a frequency band W much wider than the rate R in bit/s of the original information signal. To achieve such bandwidth expansion, a known pseudorandom sequence multiplies the information signal before the modulation takes place and it is then removed at the receiver side. Fig. 4.1 illustrates the pseudorandom signals, called chips, that, impressed on the data bits, cause the bandwidth spreading.

Resorting to a spread spectrum transmission, either in the form of frequency hopping or direct sequence, bears several benefits, namely, the robustness to different types of interference: jamming, an intentional type of disturbance; interference due to multiple

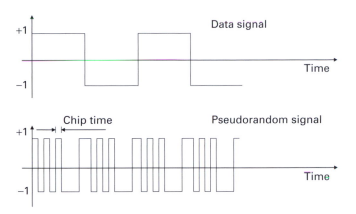

Fig. 4.1   Data stream and pseudorandom sequence in a DSSS system

access; self-interference due to multipath, i.e., to a non-linear channel that inevitably introduces distortion of the transmitted signal.

For the sake of completeness we also mention here the original IR baseband solution: it could only operate in indoor environments and relied on reflected infrared energy, without requiring a line-of-sight between emitter and receiver to work properly. The 802.11 IR standard is generally unheard, as there are no products on the market that are compliant with the IR specifications.

The FHSS standard defines a set of hopping frequency channels for transmission and reception; it also specifies a frequency channel bandwidth and provides the definition and length of the hopping patterns. Easy to understand, different sets of hopping sequences allow the co-location of similar networks in the same geographic area and also enhance the overall efficiency and throughput of each individual network.

The alternative DSSS system "chips" the baseband signal with an 11-chip pseudorandom signal. Differential binary phase shift keying (DBPSK) and differential quadrature PSK (DQPSK) modulation schemes are employed for the 1 and 2 Mbit/s native data rates, respectively.

Regarding the exact frequency channel plan, the American FCC, the Canadian IC, the Japanese MPHPT and the European ETSI standard bodies specify operation from 2.4 GHz to 2.4835 GHz, with some peculiar frequency allotments for Japan, France and Spain. The channel ID numbers and the corresponding center frequencies are shown in Table 4.1. Note that center frequencies are 5 MHz apart and that overlapping or adjacent cells can operate without mutual interference if the difference between the center frequencies is at least 30 MHz (reduced to 25 MHz in 802.11b).

The next document released, the IEEE 802.11b supplement to the original standard, specifies the high rate extension for the direct sequence spread spectrum system operating in the 2.4 GHz band. This basic new capability is called high rate direct sequence spread spectrum (HR/DSSS). The extension provides 5.5 Mbit/s and 11 Mbit/s payload data rates in addition to the 1 Mbit/s and 2 Mbit/s original rates. The higher rates are achieved through a novel modulation scheme, termed eight-chip complementary code keying

Table 4.1. Frequency channel plan for DSSS 802.11 (and 802.11b)

Channel ID	Frequency (MHz)
1	2412
2	2417
3	2422
4	2427
5	2432
6	2437
7	2442
8	2447
9	2452
10	2457
11	2462
12	2467
13	2472
14	2484

(CCK). The 802.11b document also specifies an optional coding scheme, alternative to the CCK solution, adopting a high-performing binary convolutional code, whose output is mapped to a BPSK constellation for the 5.5 Mbit/s rate, to QPSK for 11 Mbit/s. As in the previous standard document, the chipping rate is 11 MHz, and the same occupied channel bandwidth is guaranteed. The frequency channel plan is the same as in the original DSSS 802.11.

The next promulgated standard, 802.11a, describes the characteristics of a WLAN that operates in the less crowded 5 GHz band, subject to the specifications for an orthogonal frequency division multiplexing (OFDM) system with data payload communication capabilities of 6, 9, 12, 18, 24, 36, 48 and 54 Mbit/s, with mandatory support of the data rates of 6, 12 and 24 Mbit/s.

Its channels are 20 MHz wide (10 and 20 MHz in Japan) and in each channel 52 subcarriers are modulated using BPSK, QPSK, 16-QAM or 64-QAM. Forward error correction coding (convolutional coding) is employed with coding rates of 1/2, 2/3 or 3/4.

Finally, the 802.11g document, offering a higher rate extension in the 2.4 GHz band: here the PHY layer jungle appears even more crowded, as the operational modes include DSSS/CCK, OFDM as in 802.11a, plus two optional modes: one that uses DSSS for preamble and header and OFDM for payload, another which employs a single carrier modulation scheme plus a convolutional encoder. Its most salient feature is that the 6, 9, 12, 18, 24, 36, 48 and 54 Mbit/s data rates are reached, with mandatory support of the 1, 2, 5.5, 11, 6, 12 and 24 Mbit/s rates.

As the next section will clarify, the main emphasis when referring to 802.11a/b/g devices is on the possibility of building a wireless – as opposed to wired – LAN. Yet, the underlying technologies can also provide connectivity to fixed point-to-point links: the range can vary deeply, extending to several kilometers in the presence of line-of-sight propagation. On-air deployment unveils one strong limitation that the standard inevitably has to obey, i.e., the maximum allowable output power: in the 2.4 GHz frequency band it is fixed to 100 mW in Europe, 1000 mW in the USA and 10 mW/MHz in Japan; for the 5 GHz band, depending on the country and the regulatory class the single device belongs to, the maximum output power ranges from 40 mW to 1 W.

4.1.3   Architecture and MAC basic mechanisms

The main components of the 802.11 WLAN architecture are given in Figs. 4.2 and 4.3: in the left, upper part of the first figure two wireless stations – STAs from now onwards – which are able to communicate directly, form the simplest WLAN ever

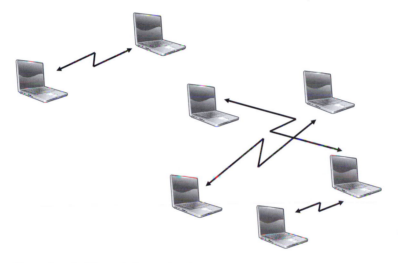

Fig. 4.2   Examples of ad-hoc wireless networks

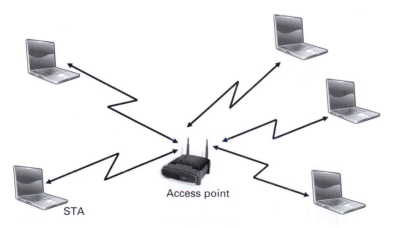

Access point

STA

Fig. 4.3   The conventional infrastructured WLAN

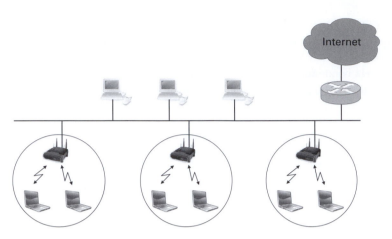

Fig. 4.4   More WLANs interconnected via a distribution system

possible: it is the smallest example, sizewise, of an independent basic service set; IBSS, as the standard documents term it.

A further example of IBSS is reported in the right, lower part of the figure, where more STAs are able to communicate directly. As this WLAN is often created without any preplanning operation, the term ad-hoc network is usually employed.

More commonly, as Fig. 4.3 exemplifies, STAs belong to a basic service set, BSS, and, unlike for the ad-hoc case, no direct colloquium is allowed among STAs: all communications are relayed via the central access point (AP). The network constituted by a single BSS is often termed *infrastructured*.

How does a single WLAN of this type fit into a more articulated network context? For residential users that set up their home WLAN, the AP is connected to the ISP network, typically via an ADSL access. For a medium size enterprise or institution, more infrastructured WLANs can be interconnected via a distribution system, as Fig. 4.4 shows.

The 802.11 MAC sublayer offers two distinct access modalities, termed distributed coordinated function (DCF) and point coordinated function (PCF): the first is for contention services, the second for contention-free services. Both can be employed in the *infrastructured* network configuration, whereas only the DCF is available in ad-hoc networks. When both are present, the corresponding periods, termed contention-free period, CFP, and contention period, CP, alternate along the time axis.

In 802.11, the main protocol ruling the radio channel access is carrier sense multiple access with collision avoidance, CSMA/CA: a member of the old, familiar CSMA family of random access techniques based on carrier sensing, it exploits an original approach to sense the wireless channel status.

Unlike in wired networks, in WLANs carrier sense is performed through the combination of a physical and of a virtual mechanism, implemented within the PHY and MAC layers, respectively. While the PHY layer simply detects the signal presence on the radio channel, the MAC layer performs a virtual carrier sense, determining whether the wireless medium is idle or busy through the request-to-send (RTS) and clear-to-send

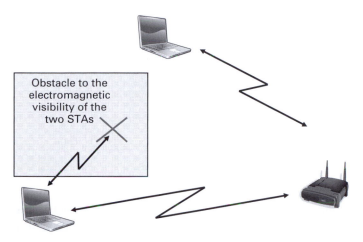

Fig. 4.5   The hidden terminal situation

(CTS) frames. In detail, before the transmission of a data frame can occur, the transmitting STA broadcasts an RTS frame and the receiving STA has to reply with a CTS frame: they both contain a duration and ID field, which specifies for how long the medium has to be reserved – hence will have to be considered busy – to transmit the data frame; the CTS frame additionally contains the corresponding acknowledgment. In this manner, all STAs within the range of the source or of the destination STA learn about the reservation.

Why isn't carrier sense performed at PHY layer enough? Because hidden terminals can be present in the BSS of the WLAN, i.e., there can be STAs that cannot hear each other's transmissions, although they all can communicate with the AP, as Fig. 4.5 indicates. Virtual carrier sense provides them with a way of knowing what is happening on the radio channel.

Every STA within the BSS collects channel status information listening to the RTS–CTS frame exchanges that appear on the radio channel and correspondingly builds what the standard calls the NAV, or *network allocation vector*: it is simply a timer, whose current value indicates when the medium will turn idle again, given the current information the STA has available.

Advantageously, the usage of the RTS–CTS frames also allows quicker inference that a collision occurred. If the return CTS frame does not reach the STA that sent the RTS frame, the STA can proceed to the retransmission more quickly than if it had to wait until the acknowledgment could be declared missing.

The RTS–CTS mechanism is not mandatory for every data frame, as its inherent overhead introduces too much inefficiency to the transmission of very short frames. It is, therefore, possible to configure a STA so that it employs the RTS–CTS frame exchange either always, or never, or only for the transmission of those data frames whose length is greater than a predefined threshold.

Regardless of whether only DCF, or both DCF and PCF are used, the first action any STA has to perform to access the channel and send a frame is to invoke the carrier sense

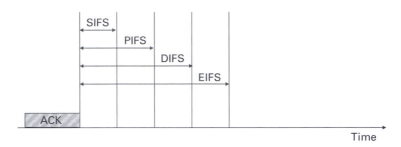

Fig. 4.6   Interframe spacing adopted by the IEEE 802.11 standard

mechanism. Unlike 802.3 MAC, where the station can transmit the frame almost immediately after the channel is sensed as idle, the 802.11 standard mandates that the medium
has to stay idle for a fixed time interval before any transmission can commence. This
interval is called the interframe space, IFS: its length varies depending on the type – and
priority – of the frame that the STA has to transmit. Moreover, the 802.11 MAC mandates
the usage of acknowledgments and retransmissions; it is another quite major departure
with respect to the MAC sublayer of wired LANs, where such notions are absent.

The IFSs the standard utilizes, listed in increasing duration order, are: short IFS, SIFS;
PCF IFS, PIFS; DCF IFS, DIFS, extended IFS, EIFS. Fig. 4.6 summarizes these different spacings: it can be immediately understood that the longer IFS the STA waits, the
lower access priority it gains. The CTS, acknowledgment (ACK) frames and fragments
of the same frame after the first are transmitted after a SIFS; the AP can gain the channel and open a contention-free interval if this stays idle for a PIFS; after a DIFS, any
station can make an attempt to transmit; the recovery of bad frames is performed after
an EIFS.

With reference to the distributed coordinated function (DCF) access mechanism, if the
sensing operation allows the STA to conclude that the medium stays idle for a time interval
greater than or equal to DIFS, then the STA can transmit its frame; otherwise it defers
transmission and keeps listening to the channel. An example of the way a data frame
and the corresponding acknowledgment are exchanged during a DCF period appears in
Fig. 4.7. The next figure, Fig. 4.8, further illustrates the RTS and CTS exchange, as well
as the corresponding NAV setting.

After deferral, or prior to attempting to transmit again after a successful transmission,
the STA waits for a DIFS plus an additional random back-off time. This extra delay
advantageously decreases the collision probability for frames that are simultaneously
ready for transmission from different STAs. The back-off time is actually a random
variable, measured in units of back-off slots: it obeys a uniform distribution, whose
lower limit is zero, whereas the upper limit is dynamically increased every time an
unsuccessful attempt to transmit occurs (i.e., when traffic builds up within the WLAN).
If the STA realizes that the medium becomes busy again during a back-off slot, then
the back-off procedure is halted and the corresponding timer is frozen until the medium
again stays idle for a time equal to DIFS: now the procedure resumes and the STA can
transmit as soon as the back-off timer goes to zero.

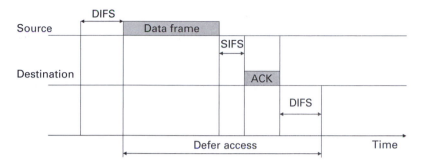

Fig. 4.7   Transmission of a data frame and of the corresponding acknowledgment during a DCF

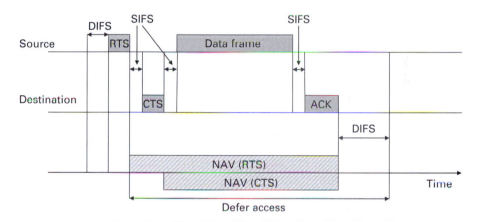

Fig. 4.8   An RTS–CTS exchange and the corresponding setting of the network allocation vector

Regarding the error recovery procedure, when the transmitting STA infers that a transmission was not successful, it is its own responsibility to handle the retransmission of the incriminated frame, until either the new attempt is successful or the maximum number of allowed retransmissions is reached.

So much for DCF. Now, let us tackle the point coordinated function (PCF). This is a centralized access mechanism: as such, it employs a point coordinator (PC), whose role is to determine which STA has the right to transmit. It is definitely not hard to conclude that PCF can be employed only in *infrastructured* WLANs, where the AP acts as the PC.

The first action performed by the PC is to take control of the channel: in turn, the PC senses the medium, and when it detects that it is idle for a time interval at least equal to PIFS – recall that the PIFS duration is shorter than the DIFS duration – it transmits a beacon frame. The contention-free period (CFP) has begun and from now on contention and contention free periods, DCF and PCF respectively, will alternate along the time axis, as Fig. 4.9 indicates.

The beacon contains the indication of the maximum CFP duration and a delivery traffic indication message (DTIM) element, which allows the frequency at which the CFP will be generated to be determined. Beacon frames are not necessarily issued at

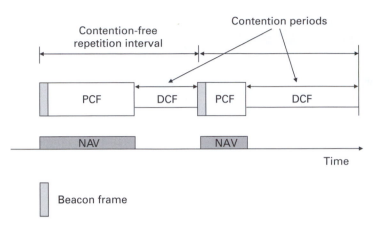

Fig. 4.9   Alternating DCF and PCF

the same frequency as the CFP: if they are more frequent, the PC will accompany the beacon transmission with the indication of the CFP residual duration. When the CFP has commenced, the PC waits for an additional SIFS, then it transmits on the medium either:

- A data frame;
- A frame to poll a STA, to enquire whether it has data to transmit;
- An acknowledgment (ACK) for a frame that the PC received right before the ACK;
- Any combination of the previous items, e.g., data + ACK, data + ACK + poll, ACK + poll, . . .
- A frame that indicates the end of the CFP.

Stations belonging to the same BSS are polled following the PC polling list order and obediently reply to the polls sending their data frames, if any. For this regulated access, no collisions occur, as under the PCF all frame transmissions use an IFS that is smaller than the DIFS.

Fig. 4.10 illustrates a possible transmission pattern during a CFP: in this example right after the CFP has been initiated, the PC sends a data frame D_1 to STA_1 and also polls this station to know whether it has data to send; after an SIFS – hence with the highest priority – STA_1 replies to the previous poll sending its data, U_1, to the PC, also acknowledging the D_1 data it just received from the PC. In turn, STA_2 is the next data recipient, D_2, and the next to be polled; in this frame exchange sequence note that the third frame appearing on the channel, although directed to STA_2, piggybacks the acknowledgment for data U_1 that STA_1 sent the PC an SIFS before; after an SIFS, STA_2 replies with data for the PC, U_2, and acknowledges D_2. Then the PC sends data D_3 to and polls STA_3, also acknowledging the receipt of U_2 from STA_2. After the next SIFS, the end of the CFP period is declared by the PC and all STAs proceed to update the corresponding NAV.

What happens if the polled STA has neither data nor acknowledgment to send? It will simply reply to the poll transmitting a NULL frame after an SIFS.

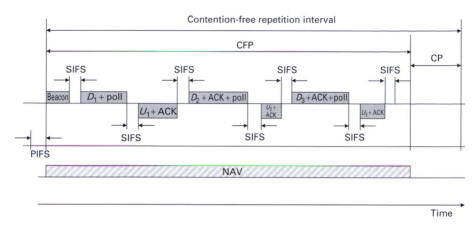

Fig. 4.10   Example of a transmission pattern during a CFP

To complete the story, observe that STAs are logically divided into CF-pollable and non-CF-pollable stations. The former STAs can reply to the PC contention-free poll and also require a poll: they are allowed to transmit a single data frame and to piggyback the acknowledgment for the data they might have received from the PC. The CF-pollable STAs and the PC do not employ the RTS–CTS mechanism during the CFP. The non-CF-pollable STAs cannot transmit data during the CFP; nevertheless, they can receive data from the PC and are required to reply sending an acknowledgment after an SIFS, just as during the contention period DCF.

Last, recall that the PHY layer of all current 802.11 implementations provides multirate support and the opportunity to switch rates dynamically, depending on transmission conditions: the rate can be increased in favorable propagation settings, decreased otherwise. Yet, the standard does not specify the algorithm that shapes rate variations: instead, it defines a set of rules that all STAs have to follow to guarantee, regardless of the dynamic rate adaptation scheme adopted, STA interoperability. This reasonably implies that no STA shall transmit a unicast frame at a rate that the receiving STA does not support. Further, all control frames (except in some particular cases), all frames directed to multicast and broadcast destination addresses, and poll frames have be transmitted at a rate that falls within the BSS-basic rate set, i.e., the set of data rates supported by *all* STAs within the BSS.

4.1.4   The need for quality of service and the 802.11e document

Wireless networks are currently used to transport a multiplicity of traffic streams: the next challenge the underlying standards have to face is to fulfil the need for differentiated Quality of Service (QoS) that distinct applications with different requirements in terms of accuracy and timeliness of delivery have. Amendment 8 to the standard [53] answers this need, introducing such differentiation within the 802.11 MAC sublayer.

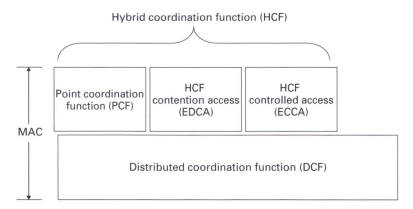

Fig. 4.11   802.11e modified architecture

The modified MAC architecture that [53] introduces is reported in Fig. 4.11: for QSTAs, i.e., for STAs supporting QoS, DCF is paired by a new function, called the hybrid coordination function (HCF), whereas for all STAs, the PCF remains optional.

In QoS network configurations, the HCF provides several enhancements, based on both a contention-based access mechanism, termed enhanced distributed channel access, EDCA, and a controlled, collision-free mechanism, referred to as the HCF controlled channel access (HCCA). The corresponding procedures let QSTAs gain transmission opportunities, TXOPs, logically termed EDCA TXOPs and HCCA TXOPs, respectively. Their duration and frequency is shaped by the QSTA needs.

Enhanced distributed channel access introduces four distinct access categories (ACs) where traffic streams with different requirements fit: in increasing priority order, they cover background, best effort, video and voice services. Within a QSTA, four distinct transmit queues, one for each AC, collect the frames to be transmitted and each AC employs a modified DCF rule to gain channel access: rather than waiting for a DIFS, every AC[i], $i = 1, 2, 3, 4$, is assigned a new interframe space, called arbitration IFS, AIFS[i], and if the channel stays idle for a time equal to the corresponding AIFS plus a conventional back-off time, the AC wins the QSTA internal contention to transmit and its frame contends externally for the wireless medium. Fig. 4.12 summarizes this modified contention-based access procedure. Access categories of the same STA with shorter AIFS are guaranteed higher access probability, and also on the common radio channel, when frames of different ACs belonging to different QSTAs contend for the medium, the same rule applies. Once the EDCA TXOP is acquired, more than one frame can be transmitted, provided that all the frames belong to the same AC.

Enhanced distributed channel access additionally employs distinct back-off times for different ACs, but this is mainly the introduction of different AIFS durations to warrant service differentiation [55].

Regarding the HCCA, its mechanism relies on the presence of a hybrid coordinator (HC), which is allotted higher priority to access the wireless medium: it therefore has to

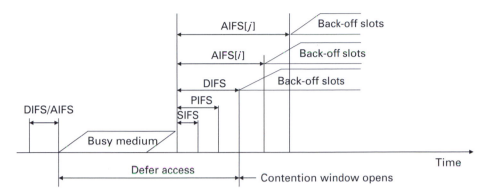

Fig. 4.12 New IFS relationships in 802.11e

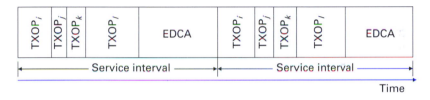

Fig. 4.13 HCCA channel access priorities

wait for a shorter time between transmissions than non-QSTAs, which use DIFS in DCF, and QSTAs, which use AIFS in EDCA: its time interval is exactly PIFS. When the HC captures the channel, it can generate a contention-free period (CFP), acting as a point coordinator; alternatively, and more interestingly, once the HC has captured the channel, it can allocate TXOPs through the HCCA mechanism to itself and to other QSTAs, which therefore gain a contention-free, time-limited opportunity to transmit their QoS frames during the contention period (CP). For HCCA, the duration of its constituent TXOPs is ruled by the specific QoS policies that the HC enforces: e.g., they can be built on some previous knowledge of the pending traffic that different QSTAs have and are not necessarily limited to a data frame duration. Fig. 4.13 illustrates a possible sequence of time intervals ruled by the HCCA.

Although successful in achieving per-category QoS differentiation, the 802.11e EDCA mechanism bundles together all voice and video packets, regardless of the stream they belong to, as there is one queue per traffic class. On the application side however, WLANs will soon face the task of delivering not only streaming media and data, but also exciting new services, such as IP-TV: enough capacity will, therefore, have to be guaranteed on very short time scales, and per-user – as opposed to per-category – requirements strictly fulfilled. Will 802.11e, with the combination of EDCA and HCCA, guarantee enough per-user throughput, not only in the best case or on average? Will it be able to avoid dropouts and warrant the seamless transport of a rich variety of applications, even when they belong to the same AC? We are all eager to know . . .

2	2	6	6	6	2	6	2	0-2304	4
Frame control	Duration ID	ADD 1	ADD 2	ADD 3	Sequence control	ADD 4	QoS control	Data	FCS

Fig. 4.14   802.11 MAC frame format

Bit 0	1	2	3	4	7	8	9	10	11	12	13	14	15
Protocol version		Type		Subtype		To DS	From DS	More frag	Retry	Pwr Mgt	More data	WEP	Order

Fig. 4.15   Frame control field

4.1.5   IEEE 802.11 frame format

Let us now take a close look at the 802.11 MAC frame format. It is given in Fig. 4.14, where, in the usual notation the length of each field appears on top, expressed in bytes. Comments on the various fields follows hereafter, preferring simplicity to thorough details, which the standard documents nevertheless offer in [49] and [53].

The first three fields (frame control, duration; ID and address 1) and the last field (FCS) constitute the minimal frame format and are present in all frames. The address 2, address 3, sequence control, address 4, QoS control and frame body fields are, on the contrary, present only in certain types of frames.

The two bytes of the first field, frame control, are organized as Fig. 4.15 shows.

The value of the protocol version is 0, all other values being reserved.

The type and subtype fields together identify the function of the frame. There are three frame types: control (bits 2 and 3 set to 01), data (10) and management (00). Each type has several additionally defined subtypes. As an example, in all management frames, bits 2 and 3 of the type field are set to 0; the beacon frame, a special management frame, corresponds to the following setting for bits 4, 5, 6 and 7 of the subtype field: 0001.

The four different combinations of next two bits, to DS and from DS, allow discrimination of different transmissions: within the same WLAN, from a STA to an external destination (via the AP), from an external destination to a STA (as before via the AP), or among APs.

The more-fragments field is set to 1 in all data or management frames that are followed by another fragment. It is set to 0 otherwise.

The retry field is set to 1 in any data or management frame that represents a retransmission of an earlier frame; it is set to 0 otherwise. It turns out to be useful at the receiving station to eliminate duplicate frames.

The power management field indicates the power management mode of the STA. A value of 1 indicates that the STA will be in power-save mode, whereas 0 indicates that the STA will be in active mode. This field is always set to 0 in frames transmitted by the

AP. As the term suggests, being in power-save mode makes quite a bit of difference for the battery lives of portable units!

When the more-data value is set to 1, it indicates to a STA in power-save mode that more data units are buffered for that STA at the AP.

The WEP field is set to 1 if the frame body field contains information that is cyphered, a notion that we will encounter again in one of the practices at the end of this chapter, when talking about security issues. The WEP field is only set to 1 within data frames and management frames, subtype authentication; it is set to 0 in all other frames.

As for the order field value, it is set to 1 in the majority of the frames, except for some non-QoS data type frames that require a specific service class.

Having completed the description of the two first bytes of any MAC 802.11 frame, let us quickly go over the duration ID field of Fig. 4.14: its contents vary with frame type and subtype, and depend on whether the frame is transmitted during the CFP, and on the QoS capabilities of the sending STA. Yet, whenever its value is lower than 32768, some of its bits give the duration value (in microseconds) of the frames transmitted during the CP, and under HCF the duration value of the frames transmitted during the CFP. The information it bears is, therefore, used to update the NAV.

Next, the address fields that the format of the frame encompasses. The four distinct addresses indicate that the approach to the addressing issue in the 802.11 standard is forcedly different from the one in IEEE 802.3 LANs: there are the source address (SA), the destination address (DA), the transmitting station address (TA) and the receiving station address (RA). To understand their usage better, it is useful to consider the following situation: in an infrastructured WLAN, a STA sending a frame to its AP, which receives it, but is not its final recipient. Three distinct addresses are necessary in this context: the address of the sending STA, acting both as the source and the transmitter, hence SA = TA; the address of the AP, which receives the frame, hence RA; the destination address, DA, which identifies the final recipient of the frame within the WLAN. If, on the contrary, the frame is sent from a STA that is outside the WLAN, the three necessary addresses are: source address, SA; AP address, TA; receiving STA, RA = DA. The four distinct addresses are all different only when there are direct links between APs, a case we choose not to explore further here.

The next field has been newly added, to cope with different QoS levels, as its name reveals. It identifies the traffic category or traffic stream that the frame belongs to and other QoS-related information. Among its different subfields, we only cite the TXOP duration requested subfield: it is an eight-bit field used to indicate the duration, in units of 32 μs, that the sending STA desires for its next TXOP for the specified traffic identifier. The range of time values is 32–8160 μs. This subfield is present in QoS data frames sent by non-AP QSTAs, with bit 4 of the QoS Control field set to 0.

The field between address 3 and address 4, termed sequence field, bears a value that indicates the sequence number of the data unit, and also indicates – if needed – the fragment number within the data unit that the frame represents. Both pieces of information remain constant in all retransmissions.

The next, familiar, frame body field carries the actual data and has a variable length.

The frame check sequence field closes the frame; as in 802.3, it contains a 32-bit CRC and is calculated over all the fields of the MAC header and the frame body field.

4.1.6   Recent enhancements: the 802.11n document

The recent 802.11e document does not – and cannot – address either the limit in the achievable transmission rates or the channel reliability issue, both inherently present in a network that operates over a wireless medium. These are the main tasks the 802.11n document draft tackles. The group working on this amendment to the original 802.11 standard is IEEE P802.11n, and began to develop the 802.11n project in 2003, to ensure the inter-operability of the next generation of WLAN devices.

The group has accepted, as a baseline, a joint proposal to amend the standard by adding physical and MAC specifications for new technologies that, at least in principle, will raise WLAN connection speeds to as much as 600 Mbit/s.

One of the main novelties of 802.11n consists of the introduction of the multiple-input–multiple-output (MIMO) technology for the air interface, i.e., of the availability to transport multiple data streams over the same frequency channel resorting to spatial multiplexing [56]. This requires multiple antennas to transmit and to receive, but guarantees a significant enhancement of both the transmission rate and the quality of the received signal.

The draft standard also indicates the use of space–time block coding; on the transmit side it optionally foresees the adoption of beam-forming antennas, sophisticated devices that can control the directionality of the radiation pattern, remarkably confining the effects of interference on the useful signal.

A further, optional factor that plays a major role in increasing the supported bit rates in 802.11n is the doubling of the communication channel bandwidth, raised from 20 to 40 MHz, with up to four spatial streams per channel: although an option in 802.11n, it is currently implemented in several 802.11n products already available on the market. Strictly speaking, these devices do not conform to a standard; rather, to a draft. Yet, the manufacturers were so eager to reach the market with these new, high performing products, that they anticipated – and not by a few months – the standard release. What if the standard final version does not reflect the draft exactly? Well, it is expected that it will be possible to upgrade the many draft-conformant products via firmware.

As for the MAC sublayer, here, too, 802.11n introduces some efficiency improvements, focusing on frame aggregation and block acknowledgments.

The former technique consists of increasing the maximum size of MAC frames, from 2304 bytes to 8000 bytes for aggregated MAC service data units, and from 2304 to 64 000 bytes for aggregated MAC protocol data units. Voice and video applications are the services that most greatly benefit from the aggregation. As a matter of fact, streams of voice traffic are made of relatively small packets, whose size depends, among other factors, on the audio codec used: collecting them in larger frames can definitely help reduce the overhead. Video sources generate bulk streams of packets and, again, the frame aggregation technique can be advantageous.

The block acknowledgment mechanism is defined in the original standard too [49], but it has never been extensively deployed; on the contrary, it is expected to be widely used in 802.11n networks [56]. Moreover, the 802.11 draft has reduced the size of the block ACK frame from 128 to 8 bytes, not a negligible saving in terms of efficiency, considering the frequency of the ACK frames.

In the metro area, 802.11n is an appealing solution for dual mode – voice over IP – terminals: not only does it increase reliability, but it also makes more capacity available to telco operators, who can therefore rely upon this technology to provide packet-switched voice services to a large number of subscribers. However, as 802.11n runs multiple OFDM streams in parallel, with a heavy chain of amplifiers, DSPs and demodulators, its battery requirements are, by far, heavier than the constraints a chipset implementing 802.11b would pose. Hence, 802.11n is more likely to be found in PC cards and gateways than in hand-held devices, where size is also a challenge. The range and coverage issues that typically affect campus and municipal wireless networks are, nevertheless, relieved by the advanced solutions of the 802.11n physical layer.

In the home environment, one promise that the 802.11n technology holds is finally to provide the necessary rates to simultaneously distribute high-definition TV and support multimedia applications in an increasing number of Wi-Fi-enabled devices. A few words of caution in this regard: the gain achieved by the spatial multiplexing that MIMO provides is very "environmental dependent." Some experiments portrayed in home environments have reported a performance that is better than the one a good 802.11g implementation achieves in half of the cases, and comparable to it in the remaining half of the locations.

4.1.7 The Wi-Fi Alliance

To provide an honest picture, the first 802.11 PC cards were often incompatible and could not communicate with access points of different manufacturers. Although late, the creation of the Wi-Fi Alliance [57], a non-profit organization, which gathers companies involved in wireless technology and services, led to certify product inter-operability through independent and rigorous testing on 802.11a, b and g devices.

More recently, the alliance introduced WMM, Wi-Fi Multimedia, a group of features for wireless networks that are based on a subset of the IEEE 802.11e QoS standard: as such, they allow the introduction of different priorities for different data streams.

Turning to the 802.11n draft standard, the Wi-Fi Alliance already offers a program to test and certify products based on IEEE draft 2.0.

4.2 Wireless MAN

Here again, quite a few IEEE standards rule the development of wireless MANs, the most important being the 802.16 document revised in 2004 [58] and the IEEE 802.16e document [59], promulgated by the IEEE 802.16 Working group in 2005.

One of the most significant aims of these standards is to specify the features of an air interface that truly provides *broadband* wireless access, as an alternative to cable, and which is successful at reducing the digital divide in areas where ADSL technology is not available.

Unlike in the original 802.11 – and most of the current 802.11 compliant products – emphasis is placed on the support of multimedia services, with QoS differentiation. Different from the 802.11 experience, the need of interoperable multivendor products is carefully addressed, and the promulgation of the standard is paired, from the very beginning, by the action of the WiMAX Forum, an industry-led, non-profit corporation that promotes and certifies compatibility and inter-operability of broadband wireless products.

The standard defines alternative PHY specifications. For frequencies below 11 GHz, where propagation without a direct line of sight has to be dealt with, the alternatives are OFDM (orthogonal frequency division multiplexing), OFDMA (orthogonal frequency division multiple access) and single-carrier modulation. For operational frequencies ranging from 10 to 66 GHz, in line-of-sight environments, single carrier modulation is envisaged.

Regarding network topology, the standard defines a primary point-to-multipoint (PMP) architecture and an optional mesh topology. In the former system multiple subscriber stations (SS) communicate with a central base station (BS), which independently handles multiple sectors simultaneously, via a sectorized antenna: a layout that closely resembles conventional 2.5G and UMTS cellular networks. Moreover, the PMP MAC is connection oriented.

In the optional mesh topology, traffic can be routed through SS and can occur directly between SSs; omnidirectional or 360° steerable antennas can be employed.

The relatively recent 802.16e document, released in February 2006, provides enhancements to support users moving at vehicular speed, specifying handover procedures to be performed between different base stations or sectors. Fixed and mobile broadband wireless access coexist in the new reference scenario, operating on licenced bands below 6 GHz.

4.2.1   Physical layer

One relevant feature of the 802.16 PHY layer is that on the downlink, adaptive modulation is used: if the propagation conditions are sufficiently benign, then a more complex modulation is employed, reverting to lower-level modulation schemes when the RF channel degrades. On the downlink, the BS supports QPSK and 16-QAM; 64-QAM is optional. On the uplink, QPSK is mandatory, 16-QAM and 64-QAM are optional; note however that it is always the BS that dictates the modulation used. Strikingly enough, adaptive modulation can occur on a burst basis, i.e., the modulation format can be varied on a time interval whose order of magnitude is the millisecond.

On the uplink, power control is also implemented, to modify the SS transmitted power depending on the channel conditions between the SS and its base.

Subscriber stations also require highly directional antennas, to minimize interference and multipaths.

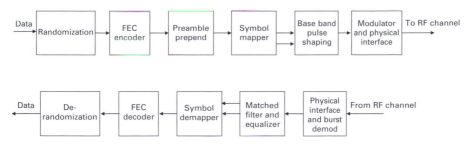

Fig. 4.16   IEEE 802.16 downlink block diagram

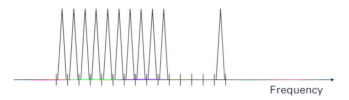

Frequency

Fig. 4.17   Synthetic representation of the spectrum of an OFDM signal

On both the downlink and the uplink, data messages are FEC encoded before being transmitted. From a physical viewpoint, the block diagram illustrating the downlink transmission is shown in Fig. 4.16, which indicates that data are randomized and FEC encoded before being fed to the symbol mapper. Control messages are also coded, but with a set of parameters known to each SS at initialization time, so that each SS can understand its content. A very similar block diagram applies to the uplink transmission, the most notable change being the absence of the equalizer in the SS.

As for the modulation, the physical layer's most interesting features are orthogonal frequency division multiplexing (OFDM) and orthogonal frequency division multiple access (OFDMA). Intentionally, the option of single carrier modulation is not further explored here. Roughly, the concept behind OFDM is to transmit multiple data symbols in parallel on a large number of separate narrowband frequency channels – subcarriers – rather than handling a serial data transmission on a single wideband frequency channel. The main advantage is to avoid complex equalization procedures, which are, on the contrary, required in the latter case. Fig. 4.17 shows a symbolic sketch of the amplitude spectrum of an OFDM signal.

In OFDMA, what is new is that subsets of subcarriers are dynamically assigned to individual users taking into account their layer-2 bandwidth needs, i.e., their data rate requirements, as Fig. 4.18 indicates, with no constraints on frequency contiguity.

To allow flexible spectrum usage, the PHY layer supports both time division duplexing (TDD) and frequency division duplexing (FDD), which guarantee bidirectional communications within the WMAN.

Regardless of the technique employed, transmissions are burst-based and data are organized into frames (once more, not to be confused with layer-2 frames). For both TDD and FDD, the allowed frame durations are 0.5, 1 and 2 ms, but the recommended

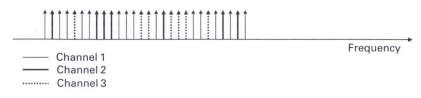

Fig. 4.18   OFDMA example

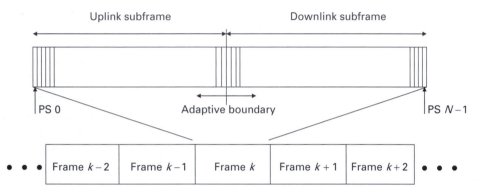

Fig. 4.19   IEEE 802.16 TDD frame structure

value is 1 ms. Each frame is made of a downlink subframe and an uplink subframe. For the TDD case they are transmitted on the same frequency bandwidth, exactly in this order, separated by a TX–RX transition gap (TTG), which in the BS allows the switching from transmit to receive mode, and vice versa in the SS; an RX–TX transition gap (RTG) analogously separates the uplink burst and the following downlink burst. The TDD frame structure is reported in Fig. 4.19, where for the first time there appears the acronym PS, which stands for physical slot and indicates a time interval whose duration depends on the symbol rate that the PHY layer adopts.

In FDD, the uplink and downlink (sub)frames are concurrently transmitted on separate frequency bands as Fig. 4.20 exemplifies. Moreover, FDD supports both full-duplex and half-duplex SSs, i.e., SSs that cannot simultaneously transmit and receive.

Getting into greater detail, the TDD downlink subframe is the one already introduced in Fig. 3.3: the subframe opens with a control section that includes a preamble for synchronization and equalization, followed by the crucial information conveyed in the uplink (UL) and downlink (DL) maps, which the BS broadcasts to all SSs: roughly, the UL map indicates the PS at which each SS can send its data, the DL map the PSs at which data bursts directed to different SSs begin. Note that on the downlink, the bandwidth available to different connections is defined with a granularity of a PS, whereas on the uplink the unit is the minislot, a time interval made of 2^m PS, with m ranging as $0, \ldots, 7$. The downlink channel descriptor DCD and the uplink channel descriptor UCD, providing information regarding the characteristics of the downlink and uplink physical channels, are also periodically transmitted in this control section, which is not encrypted. The time

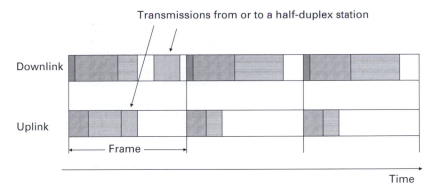

Fig. 4.20 IEEE 802.16 FDD concurrent dowlink and uplink transmissions

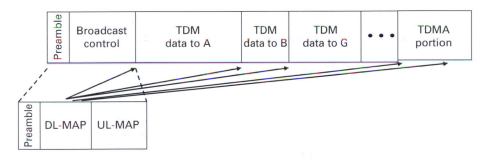

Fig. 4.21 IEEE 802.16 FDD downlink subframe

division multiplexing (TDM) portion follows, where no contention arises, as it is only the BS to broadcast data on the downlink, although to different recipient SSs.

The FDD downlink subframe displayed in Fig. 4.21 exhibits in its first portion a structure identical to the TDD downlink subframe, followed by a TDMA burst, used to transmit data to half-duplex SS (if any are present).

The uplink (sub)frame is identical for TDD and FDD. Its structure is reported in Fig. 4.22, illustrating the presence of three different periods: the first two are contention-based and are employed to transmit ranging bursts and bandwidth request bursts, respectively; the third period is contentionless and allows the transmission of bursts of scheduled data. It is left to the BS uplink scheduler to define the order and quantity of each class of bursts. Each burst in the uplink (sub)frame begins with an uplink preamble and for the case of TDD ends with a transmission time guard.

Unlike for WLAN, the 802.16 standard documents do not provide a frequency plan, as regulations in different countries have led to different spectrum allocations. Instead, what the IEEE 802.16-2004 document does is to set the RF channel sizes and some additional PHY layer features that have to be respected to ensure inter-operability over the air interface, regardless of the frequency employed in the wide 10–66 GHz range. Some of these characteristics are reported in Table 4.2.

Table 4.2. Channel sizes, baud rates and number of PSs per frame in 802.16 WMAN

Channel size (MHz)	Symbol rate (Mbaud)	Bit rate QPSK (Mbit/s)	Bit rate 16-QAM (Mbit/s)	Bit rate 64-QAM (Mbit/s)	Number of PSs per frame
20	16	32	64	96	4000
25	20	40	80	120	5000
28	22.4	44.8	89.6	134.4	5600

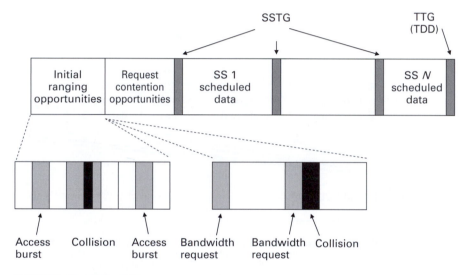

Fig. 4.22   IEEE 802.16 uplink subframe

Although many voices boost 802.16 WMAN raw rate at hundreds of Mbit/s in coverage areas of several km diameter, yet it is clear that system operation will lead to a variety of throughput values, markedly dependent on the radio channels available, on propagation conditions – LOS or NLOS, rain fades . . . – as well as on the features actually implemented in both BS and SSs.

4.2.2   MAC features

The reference model to understand the logical organization of the IEEE 802.16 standard, already anticipated in Chapter 2, is given in Fig. 4.23, which shows the convergence sublayer (CS), the common part sublayer (CPS) and the privacy sublayer. In what follows we will focus on the CS and CPS presentation.

On the transmitting side, the CS takes care of accepting data from the service access point and encapsulates them into MAC SDUs passed on to the CPS sublayer. The CS also classifies SDUs and associates them with an appropriate connection identifier (CID).

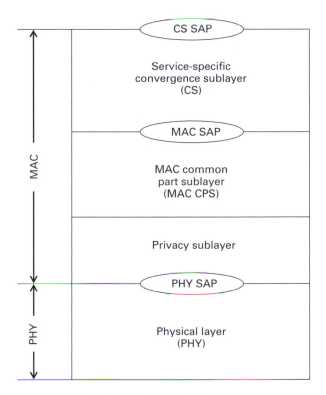

Fig. 4.23   IEEE 802.16 standard reference model

This 16-bit CID, conveyed in the generic header of any unicast frame, truly lays the basis for the QoS-aware delivery of the SDUs.

Figure 4.24 explores the notion of classification as performed by the CS at the base station. In this figure an SDU is associated with the characteristics of the service flow it belongs to, before being passed to the MAC CPS.

As for the CPS, it represents the heart of the 802.16 MAC.

In PMP mode, the common part sublayer handles the opening and closing phases of each connection and the connection maintenance and it encompasses medium access functionalities and bandwidth allocation; importantly, it schedules data transmission applying QoS criteria to each different connection.

Figure 2.7 already showed the service primitives employed in PMP mode to rule the dialogue between the CS and CPS sublayers during the connection opening phase, as well as the frames that the MAC entities actually exchange. We now enrich that description, displaying in Fig. 4.25 the semantic of a meaningful service primitive, the MAC-CREATE-CONNECTION.request. Among the different parameters passed to the requestor MAC, it is interesting to spot the scheduling service type: there are four possible choices, which we will shortly meet, and which correspond to different services, associated to more or less stringent delivery requirements; we also point out the presence of the service flow parameters, which carry indications on, e.g., peak and average data rates exhibited by the flow that the MAC connection will be opened for.

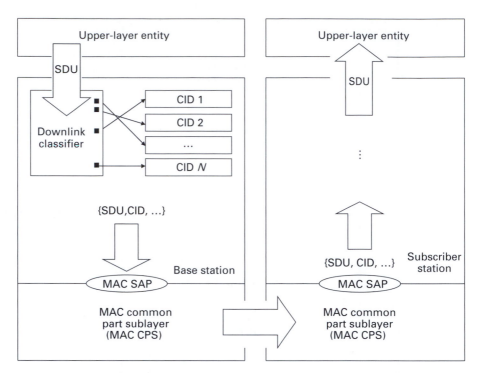

Fig. 4.24   IEEE 802.16 frame classification: from the base to the subscriber station

```
MAC_CREATE_CONNECTION.request
        (
        scheduling service type,
        convergence sublayer,
        service flow parameters,
        payload header suppression indicator,
        length indicator,
        encryption indicator,
        packing on/off indicator,
        fixed-length or variable-length SDU indicator,
        SDU length (only needed for fixed-length SDU connections),
        CRC requests,
        ARQ parameters,
        sequence number
        )
```

Fig. 4.25   IEEE 802.16 MAC-CREATE-CONNECTION service primitive

In terms of medium access control, the CPS tasks are different for the PMP case and the mesh system.

On the PMP downlink, only the base station is transmitting: it does not need to coordinate with any other station, except for the TDD case, where the channel time is divided into uplink and downlink transmission periods. On the PMP uplink, the user stations share the channel: here an articulated access protocol tailors MAC services to the delay

Table 4.3. Range of CID values and corresponding connection types

CID	Value
Initial ranging	0x0000
Basic CID	0x0001–m
Primary management CID	$m + 1$–$2m$
Transport CIDs and secondary management CIDS	$2m + 1$–0xFeFF
Multicast polling CIDS	0xFF00–0xFFFE
Broadcast CID	0xFFFF

and bandwidth requirements of each connection. Both polling, contention procedures and unsolicited bandwidth grants are envisioned, depending on the specific needs of the connections.

In PMP mode, as soon as a subscriber station SS has registered within the WMAN, the initialization phase is triggered and several connections opened between the SS and the BS. Along with transport connections, which convey data, three management connections are always opened in each direction (from the SS to the base and vice versa): they are the basic connection and the primary and secondary management connections. The basic connection is employed by BS and SS MAC entities to exchange short and urgent MAC management messages; the primary management connection is employed to send longer management messages that are more delay tolerant; the secondary management connection carries delay-tolerant, standard management messages, such as the ones generated by DHCP and SNMP.

Table 4.3 shows the correspondence between CID values and connection types; it also demonstrates that along with CIDs identifying transport and management connections, there is a connection ID used by the SS during the network entry process, CIDs explicitly reserved to send downlink broadcast messages and CIDs that are associated to groups of SSs: this latter solution will turn out to be useful when handling SS bandwidth requests. Bandwidth is granted to the SS by the BS as an aggregate in response to per-connection requests from the SS. Indeed, bandwidth requests are based on CIDs: behind a CID there might be a single session, or it may happen that more higher-layer sessions are pooled together and served by the same wireless CID. This could correspond to the situation that Fig. 4.26 exemplifies, where the WLANs act as access networks and the 802.16 WMAN is the distribution network: in this circumstance, WLAN traffic is pooled for bandwidth requests and grants.

Within the optional mesh architecture, either distributed or centralized traffic scheduling can be performed. In either case not even the mesh BS, i.e., the node that has a direct connection to backhaul services, can transmit without having to coordinate with other nodes.

When distributed scheduling is adopted, all the nodes have to coordinate their transmissions and broadcast their schedules to all neighbors, so as to avoid collisions with traffic scheduled by nodes in the proximity.

In centralized scheduling, it is the mesh BS that gathers resource requests and arbitrates the grants to transmit.

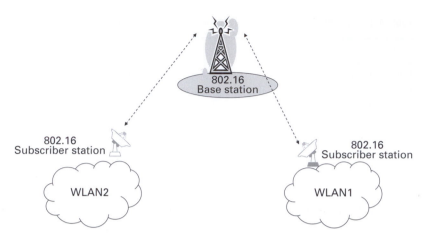

Fig. 4.26   More sessions multiplexed on a single connection

Generic MAC header	Payload (optional)	CRC (optional)

Fig. 4.27   IEEE 802.16 MAC PDU format

Within the mesh, communications occur over single links, established between two nodes.

In mesh mode the 16 bits of the CID obey a completely different syntax from that in PMP mode. However, the CID still appears in the generic MAC header of any unicast PDU sent over the mesh: within its 16 bits there are three bits that identify the priority or class of the message, so that QoS delivery can be enforced on each link of the mesh, on a per-message basis.

4.2.3   IEEE 802.16 frame format

The generic MAC PDU format is shown in Fig. 4.27: a fixed-length generic header opens the frame, followed by an optional payload. When the payload is present, it may consist of a MAC SDU or an SDU fragment, but it may also include additional subheaders. If required by a service flow, the CRC is added to each MAC PDU carrying data: in this case, a CRC, computed as in the IEEE 802.3 standard, covers the generic MAC header and the ciphered payload and is appended to the payload of the MAC PDU.

Two MAC header formats are defined. The first is the generic MAC header, the second is the bandwidth request header, which is employed to request additional bandwidth. Fig. 4.28 details the format of the generic header, and indicates the width of each constituent field in bits. The HT field is set to 0 to indicate that this is a generic MAC header; the following EC field is set to 1, meaning that encryption is used; the type field indicates the subheaders and special payload types that may be present in the payload; one

1	1	6	1	1	2	1	11	16	8
HT	EC	Type	Res	CI	EKS	Res	LEN	CID	HCS

Fig. 4.28   IEEE 802.16 generic MAC header format

reserved bit follows; the CI field is the CRC indicator: if 1, CRC is included in the PDU, otherwise it is not; EKS is the field providing the index of the traffic encryption key and initialization vector used to encrypt the payload; we then find an additional reserved bit; 11 bits for the length field, which gives the length in bytes of the MAC PDU including the MAC header and the CRC if present; finally, we encounter the CID; the header check sequence HCS field closes the header: it is an eight-bit field used to detect errors in the header.

4.2.4   Scheduling services

Let us focus on the PMP mode, where each connection is associated with a single data service, and each of these is, in turn, related to a set of QoS parameters. The four available scheduling services are termed unsolicited grant service (UGS), real-time polling service (rtPS), non-real-time polling service (nrtPS) and best-effort (BE) service.

The unsolicited grant service (UGS) has been introduced to cope with the requirements of real-time applications that periodically generate fixed-size data packets, i.e., with incompressible constant bit rate connections. Via this service, the BS guarantees fixed-size transmission grants to the SS on a real-time basis, with a period that depends on the maximum sustained rate of the service flow. Additional key information elements for the service flow are the maximum latency and the jitter that it tolerates.

The real-time polling service (rtPS) serves real-time data connections with variable-size data packets, such as video streams. Among the mandatory QoS service flow parameters we cite the minimum reserved traffic rate, the maximum sustained traffic rate and the maximum latency.

Further relaxing the requirements, we find the non-real-time polling service (nrtPS): it serves delay-tolerant applications that exhibit variable-size data packets and require a minimum data rate. Now the main mandatory QoS service flow parameters are the minimum reserved traffic rate, the maximum sustained traffic rate and the traffic priority, used to arbitrate the scheduling of different nrtPS connections.

Finally, the best-effort (BE) service, useful when no minimum service level is required: the maximum sustained traffic rate and traffic priority parameters survive to minimally detail the flow requirements. We note in passing that when asking for the BE service, the SS can use the contention request opportunities that first appeared in Fig. 4.22.

We also observe that the 802.16 MAC encompasses several mechanisms that the SS can use to convey its bandwidth request message to the BS: a stand-alone bandwidth request message, or sometimes a piggyback request. The BS can poll the SS, allocating enough bandwidth on the uplink so that the SS can respond with a bandwidth request.

Some SSs may be polled in groups or in broadcast: recall that some CIDs are reserved for multicast and broadcast messages. Moreover, SSs with active UGS connections may set a specific bit in a MAC packet of the UGS connection to notify the BS that they need to be polled.

4.2.5   WiMAX Forum

Like the Wi-Fi Alliance for 802.11, the WiMAX Forum is a corporation working to pro- mote worldwide spread deployment of 802.16 products [60]. One of its key objectives is to develop WiMAX profiles based on the IEEE 802.16 standard documents, and conse- quently assure that WiMAX-Forum-certified products are compliant and inter-operable, as they have passed the WiMAX testing and certification program. The immediate fall-off is that network operators can buy equipment from different companies and be confident that everything works together, seamlessly. But working with open standards and cer- tification also guarantees more market competition, lower prices and the possibility of addressing any inter-operability issues early, before the product is brought to the mar- ket. This is the lesson learned from the bumpy path that the Wi-Fi Alliance had to go through!

WiMAX certification testing is conducted in independent laboratories: the first were opened in Spain and Korea. In June 2008, ten certified 2.5 GHz mobile WiMAX products were the first to receive the WiMAX certified seal of approval.

4.3   WPAN: wireless personal area networks

The next shot provides a few indications regarding wireless personal area networks (WPANs). As the name suggests, the geographical scope is limited, as wireless devices of a WPAN are located a few tens of meters apart, at the very most.

This time the IEEE working group to mention is 802.15, whose work stemmed from the previous effort in developing a WPAN solution by a special interest group: the underlying wireless technology was Bluetooth, a familiar name to all of us, because it means "cable replacement" in several circumstances of daily life, mostly related to the use of cellular phones.

As for WLANs, the employed frequencies are in the 2.4 GHz band. The five channels allowed for operation are reported in Table 4.4, where the high-density and coexistence indications tell what channels would be used, depending on whether the WPAN operates in an environment where an 802.11b WLAN is also active or not. The goal is to minimize the WPAN impact on existing WLANs.

The IEEE 802.15 documents and task groups attack several issues: PHY- and MAC- layer specifications, as usual, but also the coexistence of WPANs with other wireless devices that transmit over the same unlicensed frequency bands. Moreover, low rate and high rate WPANs are distinguished, as they exhibit different requirements in terms of battery life and complexity, and target different applications, less intensive data services the former, imaging and multimedia the latter. To this regard, the IEEE 802.15.3 Task

Table 4.4. Frequency channel plan for 802.15 WPANs

Channel ID	Center frequency (MHz)	High-density	802.11b coexistence
1	2412	X	X
2	2428	X	
3	2437		X
4	2445	X	
5	2462	X	X

Group 3c is developing an alternative physical layer to the existing WPAN standard to support really high data rates; moreover, task group 5 focuses on mesh networking of WPANs, and task group 6 on body area network technologies.

As a meaningful example, we report next a succinct description of the main physical and MAC characteristics of a high data rate WPAN, thoroughly detailed in the standard document [61].

In the PHY layer, the reference data rate is 22 Mbit/s and the modulation format is uncoded DQPSK, differential quadrature phase shift keying. All WPAN devices have to support it. Other modulation formats that can be supported are QPSK, 16/32/64-QAM with trellis coding. The symbol rate for all modulations is 11 Mbaud, so that, also depending on the coding, the raw data rates supported are 11, 22, 33, 44 and 55 Mbit/s, for QPSK-TCM, DQPSK, 16-, 32- and 64-QAM TCM, respectively.

Wireless devices are organized into piconets, each piconet being supervised by one device, elected the piconet coordinator. The coordinator offers the basic timing via the beacon and manages access control to the piconet.

The channel time is divided into superframes. Each superframe begins with a beacon, followed by a contention access period and a contention-free period, termed channel time allocation period. During the contention access period, asynchronous – relatively small – amounts of data and commands, if needed, can be transmitted; isochronous streams (e.g., audio and video streaming) and asynchronous data connections are sent during the time allocation period. Here too, as in several of the standards we previously introduced, CSMA/CA is the access protocol employed for medium access during the contention period; the TDMA protocol is employed during the channel time allocation period. In the latter period the coordinating device is in charge of allocating appropriate time slots to all devices within the piconet, and can, therefore, enforce different QoS levels.

4.4   A glimpse of wireless mesh networks

To make the picture even more complicated, wireless mesh networks are now swiftly emerging. All standard groups, IEEE 802.11, 802.15 and 802.16, are actively working on the specifications for wireless mesh networks.

What features do they exhibit? In the most general sense, heterogeneity, as several wireless technologies can coexist in mesh networks.

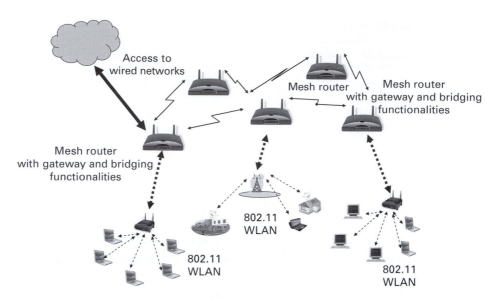

Fig. 4.29   An infrastructured wireless mesh network

From an architectural viewpoint, mesh networks can be made of mesh clients and mesh routers [62]. Mesh clients have a simpler hardware platform and software than routers, although they can also work as routers; they are typically equipped with a single wireless interface. Mesh routers, when present, form the network backbone, support multihop communications and can display multiple wireless access interfaces.

A macro classification of wireless mesh networks distinguishes between infrastructured meshing and ad-hoc networks.

The former approach is depicted in Fig. 4.29: the network backbone is provided by low mobility, wireless routers; they can connect via different radio technologies (as an example, 802.11 or 802.16), and can self-heal links among themselves; mesh clients communicate via mesh routers.

The latter solution is shown in Fig. 4.30: routing functionality is the magic new word, with respect to, e.g., 802.11 ad-hoc WLANs, as packets get to their final destination via multiple wireless hops. Also, recall that different radio technologies can be envisioned in the wireless mesh network, not necessarily 802.11, as is clearly pictured in the figure. In a pure ad-hoc network this requires dual mode radios, i.e., terminals that recognize and implement different standards in the lower layers.

In between the two extremes, there lies the hybrid architecture where some mesh clients can access the network through the infrastructure of mesh routers or via a direct mesh with other clients; this further degree of flexibility acts in favor of connectivity and coverage. Mesh routers retain the role of gateways between different networks and toward the wired Internet.

The wireless mesh network realm is still open to several challenging research problems [62]: the quest for higher bit rates has not come to an end; at the physical layer, cognitive radio, i.e., the possibility of coexistence of different systems and different

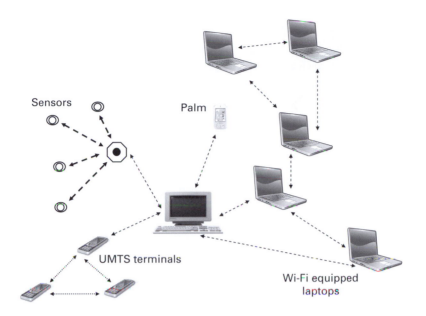

Fig. 4.30   An ad-hoc wireless mesh network

technologies in the same frequency band is a luring approach; the interaction of higher-layer protocols with the features of the physical layer, although difficult, is undoubtedly promising.

Also, how to provide the scalability of MAC protocols and how to guarantee smart bridging functions among different wireless radios still offer stimuli to novel solutions.

Efficient mesh routing is needed at the network layer, while at the transport layer adaptive rate control protocols are required, given the heterogeneity of wireless mesh networks, displaying highly variable propagation delays, transmission conditions and capacity features; even existing applications need to be improved and tailored to this turbulent context [62].

4.5   Practice: capturing 802.11 data and control frames

In this first practice we take a close look at some of the frames that 802.11 wireless stations exchange with the access point. It is relatively easy to monitor a local area network, be it wired or wireless, and to capture the frames crossing it: one useful tool available is a software network protocol analyzer called Wireshark [63], which can be freely downloaded from the Internet. In this section Wireshark has been employed once its installation procedure was completed under the Linux operating system with superuser privileges.

The setting considered in this practice for analyzing WLAN traffic is reported in Fig. 4.31 and it is the same considered in the following practices, where the management frames and the security features of recent 802.11 implementations are illustrated. The

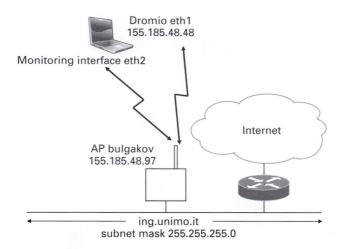

Dromio eth1
155.185.48.48

Monitoring interface eth2

AP bulgakov
155.185.48.97

Internet

ing.unimo.it
subnet mask 255.255.255.0

Fig. 4.31   Wireless setting employed for the practices

station named "dromio" has been equipped with two wireless interfaces, named *eth1* and *eth2*, the latter acting as the monitoring interface. The *eth1* NIC has been attributed the IP address 155.185.48.48; the access point is identified by the IP address 155.185.48.97, belonging to the ing.unimo.it subnetwork; the Internet can be reached via a router. The subnet mask employed is 255.255.255.0. The *eth2* NIC is configured in monitor mode on the Nth channel via the – root – linux command

```
[root@linuxbox ~]# iwconfig eth2 mode monitor channel N
```

This ensures that the captured data include not only the data frames, but also the 802.11 management and control frames.

Since Linux assigns names to devices depending on the order they are detected, we have further employed a static assignment of names to MAC addresses, so as to make sure *eth2* always identifies the monitoring interface. The Linux /etc/iftab configuration file of dromio, therefore, looks as follows:

```
# This file assigns persistent names to network interfaces.
# See iftab(5) for syntax
# Ethernet NIC
eth0 mac 00:13:d4:e3:72:3d arp 1
# wireless interface
eth1 mac 00:18:6e:2d:04:a8 arp 1
# wireless monitor interface
eth2 mac 00:18:6e:2d:04:1b arp 1
```

Let us begin by launching Wireshark and choosing it to visualize only 802.11 data frames. Once the capture interface is selected (*eth2* in our setting), one option is to collect all frames and then only visualize those that contain data, by a proper filtering rule that obeys the syntax:

```
wlan.fc.type == 2
```

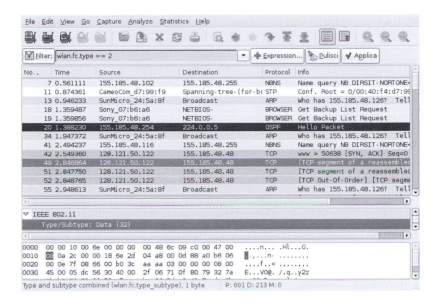

Fig. 4.32 Snapshot of Wireshark display when filtering captured data with the wlan.fc.type == 2 string

As an example, Fig. 4.32 shows the snapshot of the output that Wireshark displays when this filter is applied. In the top part of the display window there appears a list of various data frames, which Wireshark identifies via a sequence number, populating the first left column. The bottom part of the output shows the actual content of the frame highlighted in the top part.

The next figure, Fig. 4.33, provides a way to inspect the header format of an 802.11 data frame. The reader is encouraged to go through it, and spot and interpret the fields described in Section 4.1.5: it is easy to recognize that the frame indeed bears data, that the AP is sending it to a wireless station, that the source (128.121.50.122) is outside the WLAN, that the frame is a fragment and is being retransmitted. Moreover, the destination has MAC address 00:18:6e:2d:04:a8: that is no surprise, it is our only wireless station, *eth1*.

Let us now try to filter control frames: provided they are present in the captured data, the string to employ is

```
wlan.fc.type == 1
```

The specific strings

```
wlan.fc.type == 27
```

```
wlan.fc.type == 28
```

```
wlan.fc.type == 29
```

visualize the request to send frames, RTS, the clear to send frames, CTS, and the ACK frames, respectively. To successfully capture RTS–CTS frames, recall that the RTS–CTS exchange between the wireless station and the access point only occurs when data frames

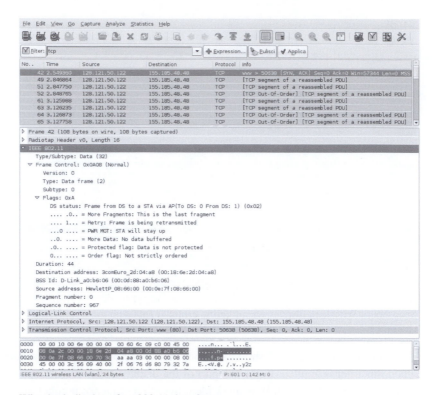

Fig. 4.33   Wireshark display of an 802.11 data frame

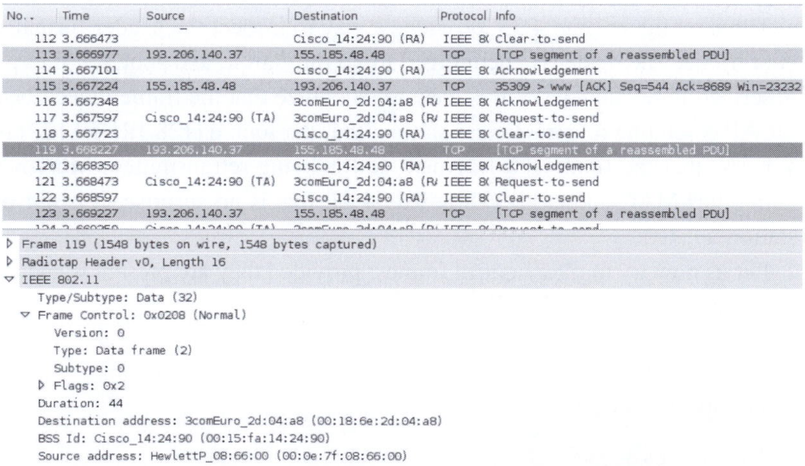

Fig. 4.34   Wireshark display of 802.11 data frames, RTS/CTS and ACK frames

reach a sufficiently robust size: in our AP this threshold is set to 1000 bytes. In Fig. 4.34, the frame with sequence number 119 – 1548 bytes in size – is transmitted by the AP after the corresponding RTS (sequence number 117) and CTS (sequence number 118) frames. The ACK follows immediately after, with sequence number 120.

4.6 Practice: inspecting 802.11 management frames

This time the intent is to see some 802.11 management frames at work: beacon frames, probe requests and probe responses, association and authentication frames and, finally, deauthentication frames. To isolate management frames, the Wireshark string to employ is:

```
wlan.fc.type == 0
```

It is interesting to start by inspecting Fig. 4.35, which provides the data field of a beacon frame transmitted by the AP: within this field it is possible to spot the time stamp, a counter that indicates how many microseconds have elapsed since the AP was turned on and that allows wireless stations to synchronize with the AP; the time interval, providing the beacon emitting period, 0.1 s in the examined case; the capability information field, bearing indications about the security features of the AP (in this circumstance the awkward wired equivalent privacy (WEP) standard was supported); and some additional elements, termed tagged parameters, which reveal the AP name, "bulgakov," the rates the AP recognizes, from 1 Mbit/s up to 54 Mbit/s, and the radio channel employed, channel 1 at frequency 2.412 GHz.

The next figure, Fig. 4.36, shows some exchanges of probe request and reply packets: recall that probe requests are broadcast by stations to understand what APs are in the proximity; they require the active participation of stations, as opposed to the passive reception of beacon frames. Notice that the second probe request shown in the figure

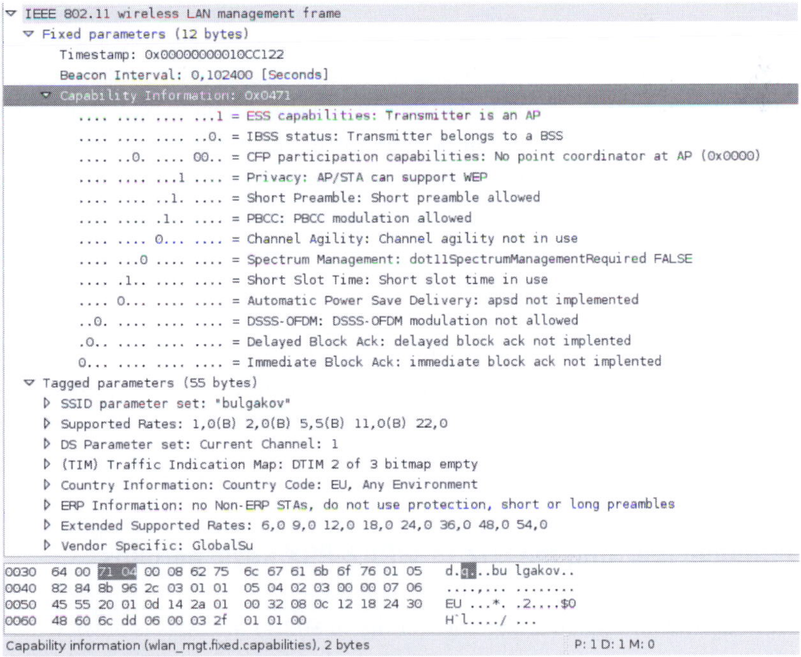

```
▽ IEEE 802.11 wireless LAN management frame
  ▽ Fixed parameters (12 bytes)
      Timestamp: 0x00000000010CC122
      Beacon Interval: 0,102400 [Seconds]
    ▽ Capability Information: 0x0471
              .... .... .... ...1 = ESS capabilities: Transmitter is an AP
              .... .... .... ..0. = IBSS status: Transmitter belongs to a BSS
              .... ..0. .... 00.. = CFP participation capabilities: No point coordinator at AP (0x0000)
              .... .... ...1 .... = Privacy: AP/STA can support WEP
              .... .... ..1. .... = Short Preamble: Short preamble allowed
              .... .... .1.. .... = PBCC: PBCC modulation allowed
              .... .... 0... .... = Channel Agility: Channel agility not in use
              .... ...0 .... .... = Spectrum Management: dot11SpectrumManagementRequired FALSE
              .... .1.. .... .... = Short Slot Time: Short slot time in use
              .... 0... .... .... = Automatic Power Save Delivery: apsd not implemented
              ..0. .... .... .... = DSSS-OFDM: DSSS-OFDM modulation not allowed
              .0.. .... .... .... = Delayed Block Ack: delayed block ack not implented
              0... .... .... .... = Immediate Block Ack: immediate block ack not implemented
  ▽ Tagged parameters (55 bytes)
    ▷ SSID parameter set: "bulgakov"
    ▷ Supported Rates: 1,0(B) 2,0(B) 5,5(B) 11,0(B) 22,0
    ▷ DS Parameter set: Current Channel: 1
    ▷ (TIM) Traffic Indication Map: DTIM 2 of 3 bitmap empty
    ▷ Country Information: Country Code: EU, Any Environment
    ▷ ERP Information: no Non-ERP STAs, do not use protection, short or long preambles
    ▷ Extended Supported Rates: 6,0 9,0 12,0 18,0 24,0 36,0 48,0 54,0
    ▷ Vendor Specific: GlobalSu
0030  64 00 71 04 00 08 62 75  6c 67 61 6b 6f 76 01 05   d..q..bu lgakov..
0040  82 84 8b 96 2c 03 01 01  05 04 02 03 00 00 07 06   ....,... ........
0050  45 55 20 01 0d 14 2a 01  00 32 08 0c 12 18 24 30   EU ...*. .2...$0
0060  48 60 6c dd 06 00 03 2f  01 01 00                  H`l..../ ...
Capability information (wlan_mgt.fixed.capabilities), 2 bytes              P: 1 D: 1 M: 0
```

Fig. 4.35   Wireshark display of an 802.11 beacon frame

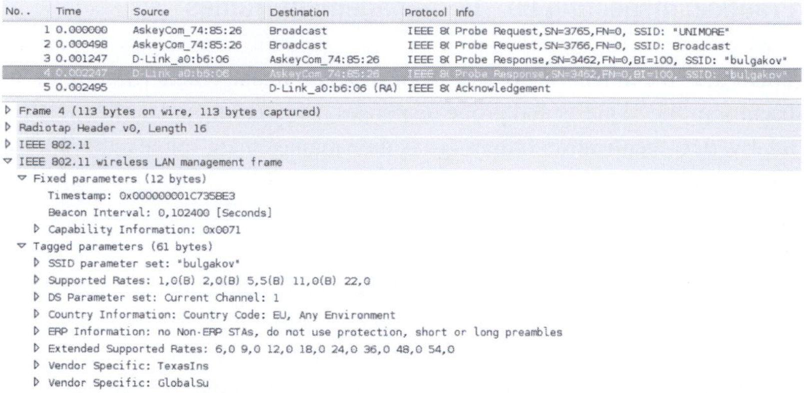

```
No. .   Time        Source             Destination        Protocol Info
     1 0.000000     AskeyCom_74:85:26  Broadcast          IEEE 80 Probe Request,SN=3765,FN=0, SSID: "UNIMORE"
     2 0.000498     AskeyCom_74:85:26  Broadcast          IEEE 80 Probe Request,SN=3766,FN=0, SSID: Broadcast
     3 0.001247     D-Link_a0:b6:06    AskeyCom_74:85:26  IEEE 80 Probe Response,SN=3462,FN=0,BI=100, SSID: "bulgakov"
     4 0.002247     D-Link_a0:b6:06    AskeyCom_74:85:26  IEEE 80 Probe Response,SN=3462,FN=0,BI=100, SSID: "bulgakov"
     5 0.002495                        D-Link_a0:b6:06 (RA) IEEE 80 Acknowledgement

▷ Frame 4 (113 bytes on wire, 113 bytes captured)
▷ Radiotap Header v0, Length 16
▷ IEEE 802.11
▽ IEEE 802.11 wireless LAN management frame
  ▽ Fixed parameters (12 bytes)
       Timestamp: 0x000000001C735BE3
       Beacon Interval: 0,102400 [Seconds]
     ▷ Capability Information: 0x0071
  ▽ Tagged parameters (61 bytes)
     ▷ SSID parameter set: "bulgakov"
     ▷ Supported Rates: 1,0(B) 2,0(B) 5,5(B) 11,0(B) 22,0
     ▷ DS Parameter set: Current Channel: 1
     ▷ Country Information: Country Code: EU, Any Environment
     ▷ ERP Information: no Non-ERP STAs, do not use protection, short or long preambles
     ▷ Extended Supported Rates: 6,0 9,0 12,0 18,0 24,0 36,0 48,0 54,0
     ▷ Vendor Specific: TexasIns
     ▷ Vendor Specific: GlobalSu
```

Fig. 4.36   Wireshark display of probe requests and responses

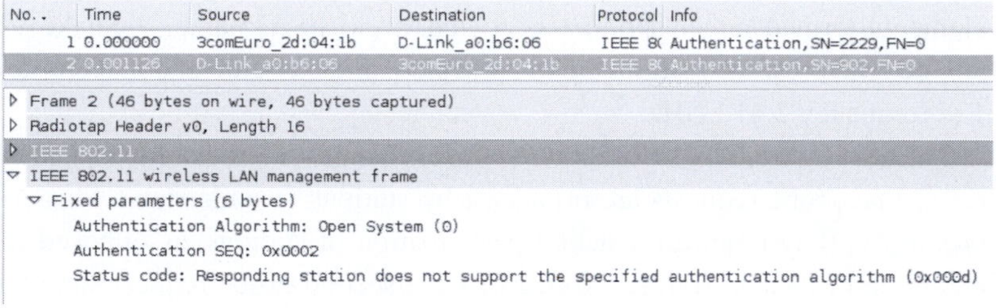

```
No. .   Time        Source             Destination        Protocol Info
     1 0.000000     3comEuro_2d:04:1b  D-Link_a0:b6:06    IEEE 80 Authentication,SN=2229,FN=0
     2 0.001126     D-Link_a0:b6:06    3comEuro_2d:04:1b  IEEE 80 Authentication,SN=902,FN=0

▷ Frame 2 (46 bytes on wire, 46 bytes captured)
▷ Radiotap Header v0, Length 16
▷ IEEE 802.11
▽ IEEE 802.11 wireless LAN management frame
  ▽ Fixed parameters (6 bytes)
       Authentication Algorithm: Open System (0)
       Authentication SEQ: 0x0002
       Status code: Responding station does not support the specified authentication algorithm (0x000d)
```

Fig. 4.37   Wireshark display of an authentication/deauthentication frame exchange

indicates "broadcast" as SSID and, therefore, triggers the probe response following immediately after, whose content is displayed in the bottom part of the figure. The keen reader should notice that this probe response has been retransmitted by the bulgakov AP, as Wireshark has captured two copies of it (with the same sequence number).

Finally, a look at a couple of authentication and deauthentication frames, the latter being sent by the AP and filtered out of the captured data with the string

```
wlan.fc.type_subtype == 12
```

Fig. 4.37 reports the content of the deauthentication frame sent by the AP to the only wireless station of the examined setting, in response to its authentication request: the setting of the security parameters between the two was intentionally mismatched before the exchange.

4.7   Practice: cracking the 802.11 WPA2-PSK keys, perhaps . . .

Although a thorough treatment of security in 802.11 WLANs is outside the scope of this chapter, this section takes a glimpse at its turbulent evolution, from the wired equivalent privacy (WEP) standard to current Wi-Fi protected access (WPA and WPA2) solutions.

The WEP standard, the original security method, aimed at guaranteeing that a WLAN would provide as much environmental privacy as a wired LAN: however, its deficiencies and pitfalls were quickly discovered and by around 2001 many programs were available to crack WEP keys, provided some cyphered packets were captured.

The WPA solution was swiftly introduced by the Wi-Fi Alliance to fill WEP gaps: it can be interpreted as a transition solution toward the security standard that the IEEE later ratified, 802.11i. Interestingly enough, it can be employed on old 802.11 devices still implementing WEP, via a firmware upgrade.

The IEEE 802.11i standard, or WPA2, as it is often termed, represents the current standard: ratified in June 2004, it relies on a multitude of additional standards, hiding behind the acronyms AES, RSNA, TKIP, RADIUS ...

Among the software tools developed to attack WLANs employing the WEP security feature in a relatively easy manner, the most popular – and instructive – are the ones within the aircrack-ng suite [64]. These programs also allow one to attempt to discover the key of a wireless network protected by WPA and WPA2, once they operate in pre-shared key (PSK) mode. In this regard, we only add that both WPA and WPA2 envisage two different operating modes: PSK and enterprise. The PSK mode requires that a pairwise master key (PMK) be shared between wireless clients and the AP; this PMK is successively employed to generate additional keys, one for each communication. The enterprise modality relies on the presence of an authentication server, which authenticates the client via additional protocols (EAP and 802.1x).

As for the WPA2-PSK attack, which we will concisely describe next, we want to appease the reader, and specify that

- Quite a bit of an effort is required to attempt to attack a current WLAN;
- More importantly, the cracking algorithms rely upon an exhaustive search and – luckily for us – in most cases they are not successful.

So, please stick to your WLAN and keep connecting through it!

The relevant programs of the aircrack-ng suite used next are *airodump-ng*, which captures the packets on a specific radio channel, *aireplay-ng*, used to force an exchange of WPA2-PSK handshake frames, and *aircrack-ng*, to perform the actual cracking of WPA2-PSK keys.

Let us see them at work in a controlled environment, where the previous AP is replaced by a newer one, with MAC address 00:15:FA:14:24:90, adopting WPA2-PSK, and let the predefined key that the AP and the wireless station share be a very naive one, "cinquecento."

The first step consists of capturing the packets exchanged between the AP and the only wireless station of the WLAN (*eth1*), invoking the command

```
airodump-ng -c 1 -d 00:15:FA:14:24:90 -w WPA eth2
```

Notice that the command is once more invoked by the root; the -c option allows one to specify what radio channel to control (channel 1); the -d option indicates to capture only frames from and to the AP with the specified MAC address; moreover, a file with prefix WPA will be created and will store the traffic sniffed by the monitoring interface (*eth2*).

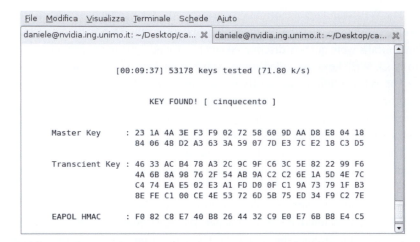

Fig. 4.38   Display of aircrack-ng successfully cracking the WPA-PSK key employed in the WLAN

Rather than waiting for a station to connect to the AP, performing the authentication handshake, it is faster to deauthenticate *eth1*, via

```
aireplay-ng -0 1 -a 00:15:FA:14:24:90 -c 00:18:6E:2D:04:A8 eth2
```

This command causes *eth2* to act as though it were the AP (MAC address 00:15:FA:14:24:90) and inject a deauthentication frame directed to *eth1* (MAC address 00:18:6E:2D:04:A8), forcing the handshake procedure to be performed again. In this manner it will be immediately possible to capture the frames we are interested in. Once the handshake is available and stored in the output file, WPA-01.cap in the example, aircrack-ng is invoked, specifying a dictionary (lower.lst is the one we used) via the following syntax

```
aircrack-ng -w lower.lst WPA-01.cap
```

The key search will commence, and for the case examined, Fig. 4.38 indicates that the cracking was successful.

The lesson learned? Never use way too simple keys, even in recent WLANs, which are WPA2-PSK protected.

4.8   Exercises

4.1   Q – Draw a histogram to illustrate the traffic load over each of the different radio channels available in the 2.4 GHz band for 802.11 WLANs between 8:30 and 11:30 a.m.

A – Hint: on each radio channel it is sufficient to count the number of 802.11 frames the monitor interface captures every, e.g., 5 minutes. Recall that it is possible to choose the radio channel to monitor with the command: iwconfig eth2 channel N

No. .	Time	Source	Destination	Protocol	Info
148	15.054108	D-Link_a0:b6:06	Broadcast	IEEE 802	Beacon frame,SN=1278,FN=0,BI=100, SSID: "bulgakov"
149	15.156488	D-Link_a0:b6:06	Broadcast	IEEE 802	Beacon frame,SN=1279,FN=0,BI=100, SSID: "bulgakov"
150	15.258863	D-Link_a0:b6:06	Broadcast	IEEE 802	Beacon frame,SN=1280,FN=0,BI=100, SSID: "bulgakov"
151	15.361237	D-Link_a0:b6:06	Broadcast	IEEE 802	Beacon frame,SN=1281,FN=0,BI=100, SSID: "bulgakov"
152	15.463734	D-Link_a0:b6:06	Broadcast	IEEE 802	Beacon frame,SN=1282,FN=0,BI=100, SSID: "bulgakov"
153	15.566111	D-Link_a0:b6:06	Broadcast	IEEE 802	Beacon frame,SN=1283,FN=0,BI=100, SSID: "bulgakov"
154	15.668484	D-Link_a0:b6:06	Broadcast	IEEE 802	Beacon frame,SN=1284,FN=0,BI=100, SSID: "bulgakov"
155	15.770862	D-Link_a0:b6:06	Broadcast	IEEE 802	Beacon frame,SN=1285,FN=0,BI=100, SSID: "bulgakov"
156	15.873227	D-Link_a0:b6:06	Broadcast	IEEE 802	Beacon frame,SN=1287,FN=0,BI=100, SSID: "bulgakov"
157	15.975737	D-Link_a0:b6:06	Broadcast	IEEE 802	Beacon frame,SN=1288,FN=0,BI=100, SSID: "bulgakov"
158	16.078103	D-Link_a0:b6:06	Broadcast	IEEE 802	Beacon frame,SN=1292,FN=0,BI=100, SSID: "bulgakov"
159	16.180476	D-Link_a0:b6:06	Broadcast	IEEE 802	Beacon frame,SN=1293,FN=0,BI=100, SSID: "bulgakov"
160	16.282858	D-Link_a0:b6:06	Broadcast	IEEE 802	Beacon frame,SN=1294,FN=0,BI=100, SSID: "bulgakov"

Fig. 4.39   Wireshark display of some beacon frames

4.2   **Q** – Once you have solved the previous exercise, configure the access point AP so as to employ the least loaded channel.

4.3   **Q** – Estimate the rate at which beacon frames are emitted by the AP.

A – *To determine the beacon rate, a capture lasting approximately 10–20 ms will suffice. Beacon frames can be filtered out using the string wlan.fc.subtype == 8 and the displayed packets saved in a file: simply count their number and divide it by the observation time interval to evaluate the required rate. In the example that Fig. 4.39 partly illustrates, 160 frames were captured in 16.28 s, so that the beacon rate is estimated as 9.83 frames/s, quite close to the 10 frames/s rate that the examined AP featured (if you do not recall it, step back to the comments on Fig. 4.35).*

```
wlan.fc.subtype == 8
```

Acknowledgments   Maria Luisa Merani wishes to thank Daniele Gratteri, one of her former students, for his invaluable help in implementing the experiments proposed in the practices of this chapter.

5 LAN devices and virtual LANs

In the previous chapters the most popular solutions for wired and wireless local area networks have been presented. Signal propagation impairments, on one side, and the multiple access technique adopted by the MAC, on the other, contribute to limit both the maximum distance that a LAN is capable of covering and its achievable performance.

However, the need to deploy LANs spanning larger areas, like an entire enterprise or university campus, has pushed toward the implementation of specific devices that allow these limits to be circumvented: repeaters, bridges and switches [65]. Such devices are also used to implement high-performance and flexible cabling schemes, according to the most up-to-date standard requirements.

This chapter describes these commonly used LAN devices, and then discusses a widely accepted solution for traffic isolation using virtual LANs. The chapter ends with a few practical examples of LAN device management and virtual LAN configuration.

5.1 Repeaters and bridges

It has already been mentioned that signal quality degrades with distance, so that a maximum span exists, beyond which a receiver cannot correctly recover the associated information.

This constraint can be bypassed using a *repeater*, an active device that links two or more LAN segments, with the goal of extending the physical connection between two LAN stations. A repeater amplifies a signal received on any input port and repeats it to all output ports. A repeater decodes incoming bit streams, re-encodes and resynchronizes them before retransmission. A repeater can then be classified as a layer-1 device. Regarding the synchronism regeneration, a repeater achieves it either by buffering part of the frame and then adding a delay, or by cutting some header bits for the time required to get resynchronized.

A physical link between two arbitrary stations in a LAN can contain a limited number of repeaters: when the number of repeaters in a chain is greater than four (remember the 5-4-3 rule in 802.3 LANs?), there is a high probability of network out-of-service. We met repeaters for Ethernet networks in Chapter 3 and observed that they also manage collisions: after detecting one, they send out the jamming sequence to all ports.

Local area networks are subject to bounds on maximum distance, load or number of connected stations. To exceed them all, it is necessary to set up an *extended* LAN. This

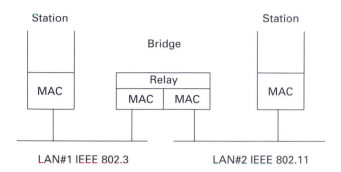

Station   Bridge   Station

LAN#1 IEEE 802.3   LAN#2 IEEE 802.11

Fig. 5.1   Bridge

LAN cannot be built with repeaters only; devices that perform functions beyond physical layer are needed and resorting to bridges is the solution.

A *bridge* is a layer-2 device that, in addition to the functionalities of a repeater, is capable of processing layer-2 data units, that is, frames. Such a device can properly read and process all fields of a frame; in particular, it understands which station is the sender and which is the destination. It can do this because it understands the layer-2 address format. As a consequence, when a bridge receives a frame it does not forward it to all other ports as a repeater does, but it stores the whole frame in a memory, it processes it and forwards it to the output along the path toward the destination only (this is what we call a *store-and-forward* operation). Equivalently, it performs a *filtering* function, meaning that a bridge forwards the frame to the appropriate outputs only, but otherwise filters it out (Fig. 5.1).

A bridge allows the number of stations in the LAN to be increased, as it guarantees each of them more layer-2 bandwidth, the actual amount depending on its processing capacity.

It is worth stressing the point that a bridge processes MAC addresses, assigned by NIC manufacturers and with no topology meaning. Nevertheless, a bridge performs simple routing algorithms and sets routing tables based on such addresses.

Most of the bridges are *transparent bridges*, a term that indicates that they host routing tables. To perform packet forwarding, they have to run a *self-learning process* to understand where the stations are placed. Moreover, they execute a spanning tree algorithm to reproduce a spanning-tree topology without closed loops.

One of the good aspects of bridges is that they are *plug-and-play* devices: they can do their job after being switched on, without requiring any programming. Other network devices, e.g., routers, cannot operate without appropriate programming and configuration.

The word *hub* is also recurrent when talking about LAN devices. Generally speaking, a hub is defined as the focal point – the star center – around which activities happen: an example of this is the hub of airlines companies. In Chapter 3 we have explicitly adopted this term to indicate multiport repeaters, but there is no consolidated, rigorous usage of it. Beginning in the years where the definition of international guidelines and standards for campus and building cabling were first introduced, "hub" generically identifies a network device to be used at the star center of a LAN topology. From the point of view of the OSI model a repeater is a layer-1 device and a bridge is a layer-2 device: in contrast, a network device that its manufacturers declare a hub can be either one, and it is necessary to take a look at its data sheets for a proper understanding of its functionalities.

5.2   Main features of bridges

The learning process is the function that allows a bridge to understand where all stations are located and how they can be reached through its output ports. When a bridge is switched on, its table is empty. As soon as it receives a frame, it forwards it to all other ports (no alternative is possible at the moment), but it also stores in the table the (sender MAC address, input port) pair. The forwarding table is updated with any frame arriving at the bridge: by inspection of the MAC source address field, the bridge learns the port to communicate with that specific station. After several data exchanges, the bridge completes the knowledge of the topology and the learning process ends. From then onwards, the bridge will forward a frame received on one port only to the port that allows to reach the intended destination.

The *spanning tree algorithm* comes after the learning process and it aims to determine a logical tree structure, deleting closed loops by disabling some output ports. It is worthwhile underlining that in case of failure this logical topology has to be updated as soon as possible with a new tree that still includes all nodes.

The spanning tree works on three steps.

(i)   Root bridge election;
(ii) Root port selection;
(iii) Designated bridge selection.

When in a LAN there are several bridges, one of them has to be chosen as the root of the spanning tree. This choice or election is usually carried on by exchanging control frames containing layer-2 addresses, as bridge identifiers. After several exchanges, the bridge with the smallest, for instance, address is elected as root of the tree. This completes step (i). Next, all other bridges decide which of the output ports has to be used to reach the root, and this represents step (ii). It may happen that a set of stations can reach the root with multiple paths through more than one bridge. In this case, it is necessary that one bridge only for the set be responsible for packet forwarding to the root, and this selected bridge is the designated bridge (step (iii)).

At the end, the LAN topology is a hierarchical tree where all stations are connected via single paths. In the case of node or link failure, the topology has to be quickly reconfigured to bypass the failure and to avoid having isolated nodes. The spanning tree algorithm is then run every time new events happen, both for station insertion and for failures.

As described in Chapter 3, modern cabling systems exhibit a hierarchical star topology and network devices are, therefore, employed at star centers. Even if a repeater might be placed in the telecommunication closet, it is more common to use a multiport bridge here, with point-to-point cables to each telecommunication outlet, to which stations connect by patch-cords. This means that if a bridge is employed as intermediate cross-connect, a hierarchical tree of bridges is explicitly given by the building cabling system, and no closed loops can occur.

5.3 Switches

Repeaters and bridges have been presented by looking at the functionalities of the OSI layer they perform. With the increase of computing power of workstations and personal computers and of data processing of NICs on the one hand, and with the evolution of building cabling systems on the other, it becomes crucial to design LANs with no bottlenecks for data flows between source and destination.

In particular, a multiport bridge placed as star center in a LAN topology, in the telecommunication closet, as intermediate or as main cross-connect does not have to represent the bottleneck for data flows entering its ports. A *switch* can then be seen as a multiport bridge designed and developed to avoid bandwidth bottlenecks in the LAN. It has a lot of network cards, one per port, which directly connect hosts to the network. A switch must then be equipped with processors and memories capable of supporting and processing multiple parallel data flows. If an Ethernet switch is taken into account, every port with its card and with the connected host is actually a collision domain in which all the bandwidth is given to the host. If the internal hardware is such to support all incoming traffic, every host can work at full rate in full-duplex mode. For instance, a switch can have 16 fast Ethernet ports and one gigabit Ethernet uplink port so that many parallel flows can be multiplied into the uplink port while others can be switched between couples of ports, at the same time.

Switches are usually of two types, *store-and-forward* and *cut-through*. While the former is similar to the one seen for bridges, the latter allows for incoming packets to be switched, forwarded and retransmitted without necessarily being completely received: the head of a packet can then be transmitted onto the addressed output port while the tail has not yet been received. Of course, this can only be possible if the hardware has enough data processing capacity and it allows for drastically reducing the latency time, typically from ms to μs.

5.4 Virtual LAN

Since their birth, and through their evolution, *virtual local area networks* (VLAN) have represented one of the most powerful and flexible solutions to build switched environments: their most evident advantage is to allow the logical grouping of stations into disjoint broadcast domains, regardless of their physical location. In the past, their widespread deployment has been limited by the lack of full compatibility between different brands of equipment. Fortunately, things are changing in current years, owing to the IEEE releasing the 802.1Q standard [66], which defines an architecture for virtual bridged LANs and their services, as well as the protocols and algorithms that allow the provisioning of such services.

5.5   Overview: VLAN definition and benefits

Let us begin with a qualitative definition of a VLAN, affirming that a VLAN is a set of logically grouped stations, subject to no restrictions on their physical location. Unlike traditional LANs, where the wiring of the buildings does sometimes limit the capabilities for rearranging, inserting and removing stations, VLANs effectively bypass the issue. Virtual LANs do not require that a group of stations sharing some intents, such as all R&D workstations and R&D LAN-connected devices, is connected to the same switch and then very probably lies within the same area. The concept is better illustrated in Fig. 5.2, reporting how a set of stations may be grouped in a traditional (LAN) manner, or in a more flexible (VLAN) fashion. In the first case, stations are usually connected based on both location and activity constraints: e.g., all R&D workstations could be located on the same floor and then be exclusively connected to the bottom left switch. In the latter case, the second constraint vanishes and stations are solely grouped according to their activity, regardless of where they are and of what switch they are connected to. As VLANs are software-based, there is no need to move equipment or cables to perform VLAN creation and maintenance.

Getting more technical, we state that the key feature a VLAN displays is that all its stations share a unique broadcast domain. This statement is a rigorous definition for the term VLAN. To enforce it and make the definition even clearer, we observe that distinct, non-overlapping broadcast domains correspond to different VLANs. Two immediate and significant implications arise from the definition just provided. On network performance: confining broadcast traffic, which is often one of the factors contributing to performance degradation, helps in saving bandwidth. On security: separate broadcast domains set natural constraints on inbound and outbound VLAN traffic. One more advantage of VLANs is improved flexibility: think of a mobile user who unplugs his laptop from a switch and plugs it back into a second switch located on a different floor: good VLAN

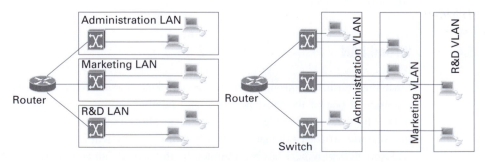

Fig. 5.2   Devices grouped in a traditional manner (LAN) or more flexibly connected (VLAN)

VLAN major benefits	
Flexibility	Stations are logically grouped, rather than being assigned a placement based on physical constraints (existing cabling, presence of switches...)
Simplified network management	As a result, network administrators can manage VLANs more easily than traditional LANs: moving stations from one place to another, removing and adding stations are greatly simplified actions.
Inter-operability	All IEEE 802 LANs support VLAN
Performance and security	As each VLAN is a broadcast domain, broadcast frames from other VLANs are filtered out; this limits the performance degradation due to unnecessary broadcast traffic and calls for explicit traffic rules for inter-VLAN communication.
Equipment savings	In traditional LANs, a concentrator (hub, switch) only serves a group of users and services. In VLANs, a single switch may theoretically serve as many LANs as its ports.

Fig. 5.3   Major VLAN benefits

administration allows the laptop VLAN membership to stay the same, despite the fact that the laptop location has changed. A facilitated management for network administrators is another strong point in favor of VLANs. Finally, VLANs are supported over all IEEE 802 LAN MAC protocols, rather than as a secondary aspect when dealing with wired–wireless networks, given the proliferation of 802.11 hot-spots! They are also supported over shared media LANs as well as point-to-point LANs. The major benefits of VLANs are summarized in Fig. 5.3.

Concerning drawbacks, very few can honestly be foreseen: a minimum additional transit delay, a negligible increase in the frame loss and duplication rates introduced by the switches participating into the VLAN.

5.6   VLAN classification

Deploying virtual LANs can be either a straightforward or complex task depending on specific needs, functions and number of managed stations. A VLAN may be limited to a single switch in small environments, or span multiple switches whenever the number of stations requires a branched architecture. Often, VLAN types are classified with respect to the nature of the association between user memberships and the VLAN itself: a distinction is, therefore, possible between static and dynamic VLANs. In *static* VLANs, the network administrator assigns switch ports to a specific VLAN. Once a switch port is configured to belong to a certain VLAN, the corresponding association holds statically: the port will not be associated with a different VLAN unless the administrator deliberately changes the configuration. In other words, rather than a station owning a membership to a VLAN, it is the switch port itself that is permanently associated with a VLAN.

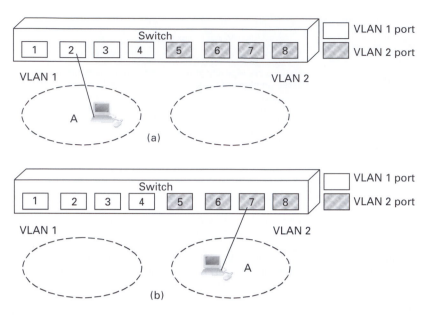

Fig. 5.4   User migration between static VLANs

This explains why static VLANs are sometimes referred to as port-based VLANs. For example, ports 1 through 4 of a switch can be assigned to VLAN1, while ports 5 through 8 are part of VLAN2. A station, such as a laptop computer, will acquire VLAN1 or VLAN2 membership depending on the port it connects to. A very simple example of migration between static VLANs is illustrated in Fig. 5.4. Station A is initially a member of VLAN1 (a), since A connects to the second port of the switch. If A leaves VLAN1 (b) and plugs into an unused port assigned to VLAN2, say port 7, from that moment on station A is not able to communicate with VLAN1 stations any longer, as it happens to fall in a different broadcast domain, virtually on a different LAN. Indeed, this is a static configuration!

In a *dynamic* VLAN, the criterion leading the VLAN membership is the station address, either layer-2 (MAC address), or layer-3 (network address), rather than the port. To be able to associate a station with the VLAN it belongs to, a switch has to maintain a VLAN database that binds the station address to the proper VLAN. The most deployed dynamic VLAN type is MAC-based and it is to this choice we will refer from now on; dynamic VLANs based on network addressing are not common, as the widespread adoption of DHCP services limits their effectiveness. Dynamic VLANs are based on specific software embedded into the switch.

Figure 5.5 shows the case of station A dynamically assigned to VLAN1. In (a), A is connected to port 2, while in (b) it is connected to port 7. Unlike in the static VLAN case, in case (b) user A inherits VLAN1 membership, as the switch binds the MAC address of the NIC inside the station to VLAN1 through its VLAN database.

Easy as it might sound, this MAC address-to-VLAN mapping hides a higher and more subtle complexity: how does the switch know through what port it can reach each MAC address, hence NIC, hence station? We already possess the right answer: dynamically. Recall that the learning process already described in this chapter with reference to bridges

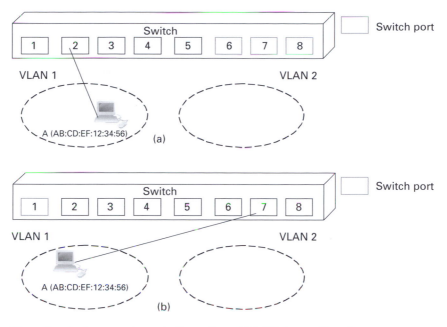

Fig. 5.5   User migration between dynamically configured VLANs is not allowed

indicates that the bridge ignores how to reach a station until it receives on one of its ports a frame carrying in its source address field the MAC address of the station. It is on this event that its switching database is created or updated. Such a database, paired with the VLAN database, allows the switch to handle dynamic VLANs properly. The drawback is that unnecessary broadcasts can occur until the switching database construction is not complete, and whenever a station is inserted, moved or rearranged.

Which of static and dynamic VLANs is better? The answer to this question requires an explanation, rather than just a crude verdict. There are several factors that affect the decision to deploy a static or a dynamic VLAN; two crucial aspects are the number of stations to be grouped and what their activities are about. If the number of stations is low, static virtual LANs are a feasible solution, as they are easy to configure and to manage. If stations are seldom moved from their position and network physical topology is not often subjected to major changes, then static VLANs are also very practical. In large environments, where frequent changes occur and flexibility in dealing with user mobility is essential, then dynamic VLANs may represent a more adequate and scalable solution. In general, organizations tend to avoid deploying dynamic VLANs, as they are difficult to administer, usually spanning multiple switches. Dynamic VLANs, which are explicitly based on software, usually exhibit a slower response time than static LANs.

5.7   VLAN on a single switch

The most intuitive way to step-by-step introduce VLANs and elaborate more on their features is to start reasoning on a single switch architecture: for simplicity, let us assume

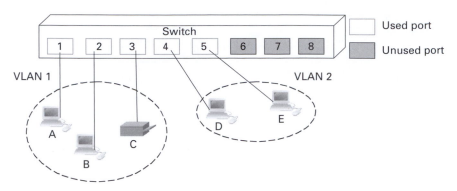

Fig. 5.6   Two VLAN domains configured on a single switch

that behind each port of the switch hides a single station equipped with a NIC, as depicted in Fig. 5.6. The first problem to solve is to create distinct VLANs (say two) within this group of stations, in order to have stations A through C belong to VLAN1 and stations D and E belong to VLAN2.

When the static configuration is adopted, the solution is to declare that the switch ports that stations A through C are connected to (ports 1 through 3) belong to VLAN1, whereas the switch ports of stations D and E (ports 4 and 5) belong to VLAN2. This is achieved through a proper switch configuration performed by the network administrator: it is a process that normally encompasses two steps, first the creation of two VLANs with their respective VLAN identifier, VID, then the assignment of the desired switch ports to each VLAN. As a result, the VLAN database is created: each of its entries specifies a VID-port binding. Once the switch is configured, whenever the switch receives a frame from a port of VLAN1(2), it will broadcast the frame only to those ports associated to VLAN1(2). The goal of creating two VLANs has been primitively reached.

More flexibly, we can opt for the dynamic configuration of the desired VLANs. In our simplified scenario this translates into providing the switch with (i) the correspondence between VLAN1 and the list of MAC addresses for stations A through C; (ii) the correspondence between VLAN2 and the MAC addresses of stations D and E. In essence, we are requiring canonical management operations, conceptually not dissimilar to the ones the static configuration requires. The add-on is that if station A connects to port 6, it is still recognized as a VLAN1 member.

For both the static and dynamic configuration, the final result is to have implemented – via software – two separate switches in the original single switch. The two switches are not aware of each other's presence, and intentionally, they cannot exchange information: any station on VLAN1(2) is both precluded transmission to, and reception from, any other station on VLAN2(1).

We close this section by noting that in the very simple scenario we have so far examined, handling VIDs is an exclusive competence of the switch: the only location where VIDs appear is the VLAN database. Virtual LAN IDs are a completely extraneous concept to NICs, which keep generating frames according to the MAC standard they conform to; the switch works on old, familiar frames.

5.8 VLAN on multiple switches

5.8.1 The need for tagging and virtual topology

More realistically than in the manner exemplified in the previous section, a VLAN is expected to span onto a more complex topology than the one a single switch represents, i.e., onto a switched LAN: we name this type of VLAN virtual switched LAN.

Let us, however, discriminate the issue of creating a VLAN over a switched LAN of moderate size from the one of creating – and, most importantly, maintaining – a VLAN over a medium to large-sized switched LAN.

To begin with, consider a small switched LAN, like the one in Fig. 5.7: stations belonging to different VLANs are physically connected to different switches; a switch-to-switch connection allows intra-VLAN communication. The two points worthy of mention are:

1. Frames originating within both VLANs need to traverse the connection between the two switches;
2. The two switches will have to share some type of VLAN database, no matter whether a static or dynamic VLAN configuration is adopted.

Let us now make an effort to understand the rationale behind the actual solution devised to build the desired VLANs. If we extended the reasoning illustrated in the previous section for static VLANs, such a brute force approach would require each switch to know the assignment of the other switch ports for the shared VLAN; even worse, for dynamic VLANs each switch should know all the MAC addresses of the shared VLAN members. What would happen in a more general setting, like the one suggested by Fig. 5.8, where the necessity is to have many broadcast domains, spanning many

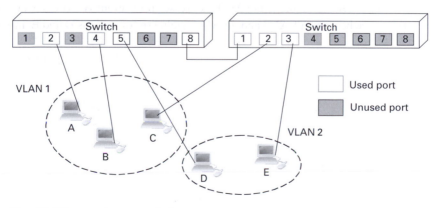

Fig. 5.7   Two VLANs spanning two switches

switches? Would complex VLAN databases and the consequent burdensome switch-forwarding operations represent the correct answer? Clearly not. Indeed, a smart, scalable design cannot be achieved in this manner.

The solution is partly provided by the IEEE 802.1Q standard. The magic word is: *tag*. Let each frame carry a tag, let the tag contain a VID and let the switches exclusively rely on VIDs for inter-switch forwarding. Indeed, tag introduction does facilitate VLAN database handling and switch forwarding. Moral: in Fig. 5.7 we will see a differently tagged frame that only circulates over the switch-to-switch connection.

On the other hand, tagging a frame requires an ad-hoc frame format, with respect to the one that, e.g., 802.3 and 802.11 MAC employ: at least one new, additional field is needed to carry the VID. Necessarily (and luckily) only switches are responsible for tag insertion and removal, and know how to interpret the tag content.

We end the section by introducing three logically distinct views, helpful at describing the challenging scenario of virtual switched LANs exemplified in Fig. 5.8. We begin by distinguishing the physical topology, which encompasses stations, hubs, switches and connections, from the active topology that the switched LAN exhibits: depending on whether the switched LAN adopts the spanning tree protocol or not, the active topology can or cannot coincide with the physical topology. We then complement these concepts with the virtual topology view: this further abstraction level is nothing but a clean picture of the VLANs we want to implement on top of the switched LAN, fully described by its physical and active topology. In conclusion, we state that the general issue is to create and manage the virtual topology layout (i.e., VLANs) on top of the active topology, as summarized in Fig. 5.9.

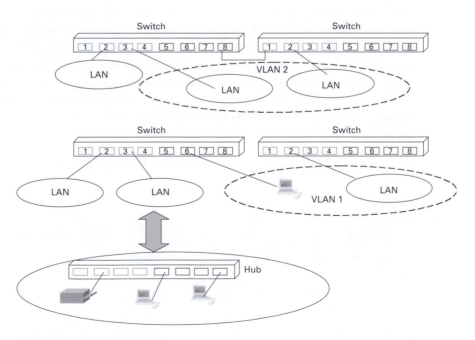

Fig. 5.8   Multiple-switch VLAN scenario

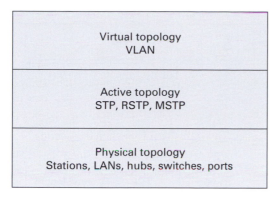

Fig. 5.9   VLAN overview

Fig. 5.10   Ethernet-encoded tag header for Ethernet and 802.3 MAC methods

Would the static configuration be appealing and competitive with the dynamic one in virtual switched LAN? If reliability is a concern, we answer by observing that dynamic VLANs are indeed more complex to administer, but if the physical topology allows for alternate, redundant connectivity to LANs, they are preferable, as more robust with respect to network component failures.

5.8.2   IEEE 802.1Q frame tagging

As we anticipated, tagging of frames needs to be performed to allow a frame to carry a VID in virtual switched LANs. To be accurate, we should, however, mention that tagging, ruled by some parts of the IEEE 802.1 standard, is introduced for the following additional purposes:

- To assign user priority information to frames carried on IEEE 802 LAN MAC types with no inherent ability to handle priorities at the MAC layer (802.3 and Ethernet methods are the example to cite with reference to this point);
- To allow VLANs to be supported across different LAN types;
- To allow the frame to indicate the format of MAC address information.

Intentionally, we will confine the description of the tagged frame structure to the format that 802.3 and Ethernet methods adopt.

To tag a frame means to add a tag header to it, immediately following the destination and source MAC address fields and, as an immediate consequence, to recompute the frame check sequence (FCS). The general tag header form for 802.3 and Ethernet MAC methods is the one reported in Fig. 5.10, including the tag protocol identifier, TPID, the

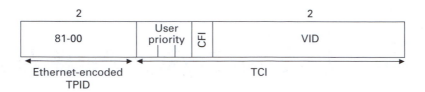

Fig. 5.11   TPID and TCI format

Destination MAC address	Source MAC address	81-00	TCI	Type/length
6	6	2	2	2

Fig. 5.12   First 18 bytes of a tagged Ethernet or 802.3 frame

tag control information, TCI, and the embedded routing information field, E-RIF. We will not spoil the amusement the interested reader will experience when consulting the faultless TPID, TCI and E-RIF explanations that the IEEE 802.1Q standard documents provide. Instead, relying on Fig. 5.11, we will comment the TPID and TCI formats and values typically encountered. In strict order of appearance:

1. The TPID field: two bytes for Ethernet or 802.3 frames, it takes on the 81-00 value, which represents the type value assigned to 802.1Q; such an unambiguous choice lets switches understand that this is neither an admissible value of the Ethernet-type field nor a value that the 802.3 length field might show. On its reception, the switch has to get ready to interpret the TCI field that follows correctly.
2. The TCI field, encompassing:

    a. The user priority field: three bytes in length, interpreted as a binary number, it is capable of representing eight distinct priority levels, 0 through 7;
    b. The canonical format indicator CFI, a single bit flag value: when reset (as it normally is) it pleasingly indicates that the E-RIF field is not present;
    c. The twelve-bit VLAN identifier VID field: it uniquely identifies the VLAN that the frame belongs to. A VLAN-tagged frame contains a VID value other than the null and the reserved FFF values. A priority-tagged frame has a VID value of 0.

In summary, the first eighteen bytes of a tagged frame are expected to look as in Fig. 5.12.

5.9   Inter-VLAN communications

As we have stressed throughout the entire chapter, a VLAN creates a broadcast domain. Therefore, the presence of a layer-3 device – a router – is mandatory to allow members of different VLANs to interact.

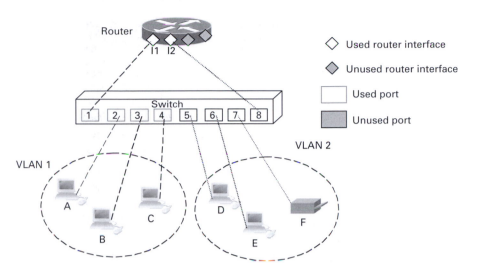

Fig. 5.13 Inter-VLAN connectivity on a single switch and with multiple physical links

The now familiar intra-VLAN communication issue is conceptually different from the inter-VLAN communication problem. In the former case, stations wishing to exchange information must be members of the same VLAN, regardless of the VLAN spanning a single switch, or multiple switches. In the latter, stations that are already members of different VLANs wish to interact: even if all VLANs were confined on a single switch, nevertheless they would have to resort to a router to let this happen.

For the sake of simplicity, we describe different alternatives for inter-VLAN connectivity limiting the discussion to single switch architectures. A simple scenario is depicted in Fig. 5.13. Suppose that we wish to deal with static VLANs, equally sized, with ports 1 through 4 reserved for VLAN1 memberships, and ports 5 through 8 assigned for use within VLAN2. Next, assume that VLAN1 members have to print documents using printer F located within VLAN 2. As stated so far, this action cannot be performed without employing a routing device. In the presented scenario, physical links – one for each VLAN – are used to offer connectivity between VLAN1 and VLAN2. Hence, the switch is connected to the I1 and the I2 router interfaces through ports 1 (VLAN1 side) and 8 (VLAN2 side). At this point, if VLAN1 stations need to print over F, the router can switch the traffic from interface I1 to interface I2, as it normally does when routing packets.

The discussed solution has the advantage of being very easy to implement. On the other hand, it lacks scalability: if the network grows in size, switch ports and router interfaces will rapidly be consumed in switches-to-router connectivity. In the example, a maximum of six ports out of eight can be employed for stations, and two router interfaces out of four serve just a pair of small-sized VLANs.

A more intelligent approach to route inter-VLAN traffic is to use logical, rather than physical, connections. This scenario is shown in Fig. 5.14.

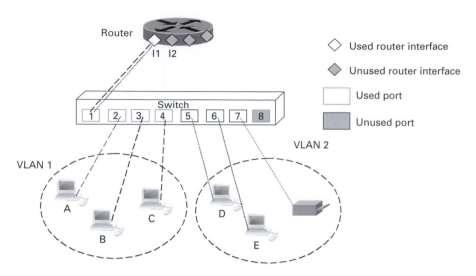

Fig. 5.14   Inter-VLAN connectivity on a single switch and with a single physical link

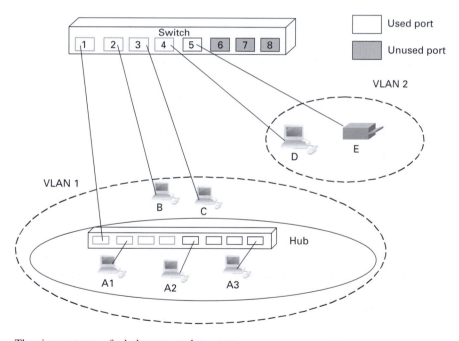

Fig. 5.15   The circumstance of a hub connected to a port

The corresponding solution adopts logical links – one for each VLAN – grouped into a single physical link that connects one switch port to one router interface. Logical links are created using specific software embedded in the routing and in the switching devices that supports the splitting of a single interface (port) into multiple logical interfaces (ports). Such an operation is painless to perform, and results in a single physical interface (port) acting like several physically independent interfaces (ports). The virtual interfaces created are usually referred to as sub-interfaces. When VLAN1 and VLAN2 members need to communicate, the traffic between the switch and the router from both VLANs exploits just one switch port (port 1) and one router interface (I1). With respect to the previous scenario, we save one switch port and three router interfaces.

Using logical connections for inter-VLAN connectivity adds minimal overhead to the implementation process, and is a very powerful resource, as it implies saving physical resources and limiting costs. Scalability is another great benefit of such an approach.

5.10 Practice: switch management and VLAN configuration

While small switches used in SOHO environments provide only the basic layer-2 switched-Ethernet functionalities, medium- to large-sized switches used to build enter- prise LANs are equipped with management software that allows to improve network performance through VLAN configuration, spanning tree protocol implementation and remote monitoring and management. This section provides practical examples on how to manage Ethernet switches and configure virtual LANs. The following tools have been used:

- An Ethernet switch equipped with management software;
- A Linux box with 802.1Q VLAN support;
- A terminal emulator;
- A null-modem serial cable;
- Three TCP/IP hosts with Ethernet NICs.

5.10.1 Switch management

Ethernet switches equipped with management software typically provide three different management methods: a traditional command line interface (CLI), a standard SNMP interface and a user-friendly web-based interface. The CLI can be accessed either through telnet/ssh clients via one of the Ethernet ports or by connecting a terminal emulator to the switch serial port using a null-modem cable. The other two management interfaces can be accessed via the Ethernet ports using either a generic web browser or a standard SNMP network management application – for further details on the simple network management protocol (SNMP) see RFC 3411.

To access the CLI via the serial port, the *kermit* terminal emulator can be used on a Linux box, as shown by the following lines:

```
[root@linuxbox ~]# kermit
C-Kermit 8.0.211, 10 Apr 2004, for Red Hat Linux 8.0
 Copyright (C) 1985, 2004,
  Trustees of Columbia University in the City of New York.
Type ? or HELP for help.
(/root/) C-Kermit>set modem type none
(/root/) C-Kermit>set line /dev/ttyS0
```

```
(/root/) C-Kermit>set carrier-watch off
(/root/) C-Kermit>set speed 19200
/dev/ttyS0, 19200 bps
(/root/) C-Kermit>set parity none
(/root/) C-Kermit>set stop-bits 1
(/root/) C-Kermit>connect
Connecting to /dev/ttyS0, speed 19200
 Escape character: Ctrl-\ (ASCII 28, FS): enabled
Type the escape character followed by C to get back,
or followed by ? to see other options.
----------------------------------------------------
Login:
```

Once the terminal emulator has been launched, a few parameters must be set, according to the specifications of the switch serial interface. In our example, a null-modem is specified, `ttyS0` is the serial port to be used on the Linux box, no carrier must be expected, the baud-rate is set to 19200 bps, no parity check is performed and a single stop bit is used. Then the `connect` command activates the communication and a login mask is shown.

After a successful login, a text-mode menu is shown similar to the following:

```
Menu options: ------------------------------------------------
  bridge                - Administer bridging/VLANS
  ethernet              - Administer Ethernet ports
  feature               - Administer system features
  ip                    - Administer IP
  logout                - Logout of the Command Line Interface
  snmp                  - Administer SNMP
  system                - Administer system-level functions

Type ? for help.
-----------------------------------------------------------
Select menu option:
```

The available options are self-explanatory. In particular, the `ip` option allows an IP address to be assigned to the switch. This is required to be able to use the web and SNMP interfaces. Figure 5.16 shows the switch IP configuration visualized through the web-based interface.

Assuming that four hosts are connected to switch ports 1, 3, 8 and 10, as shown in Fig. 5.17, it is possible to visualize the switch database content, including the MAC address and port number of each machine connected and known to the switch. The following is the CLI visualization, while the web-based version is shown in Fig. 5.18.

```
-----------------------------------------------------------
Select menu option (bridge/port): address

Menu options: ---------------------------------------------
  add                    - Add a statically configured address
  find                   - Find an address
```

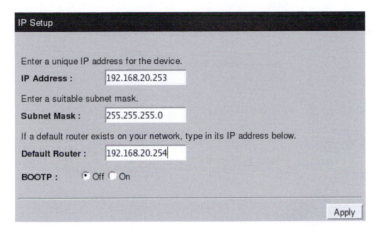

Fig. 5.16   Switch IP configuration through web-based management interface

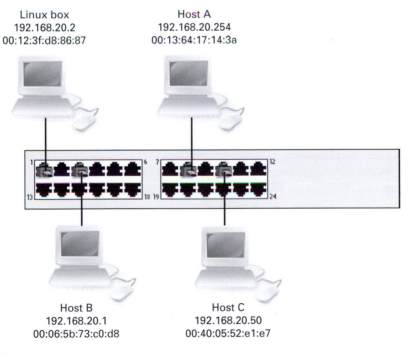

Fig. 5.17   Experiment set-up

```
    list              - List addresses
    remove            - Remove an address

Type "q" to return to the previous menu or ? for help.
--------------------------------------------------------
Select menu option (bridge/port/address): list
Select bridge ports (1-24, all): all
```

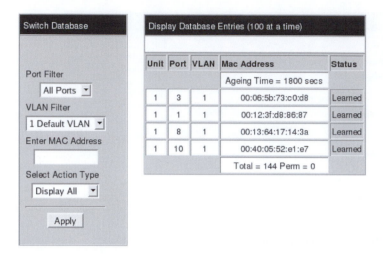

Unit	Port	VLAN	Mac Address	Status
			Ageing Time = 1800 secs	
1	3	1	00:06:5b:73:c0:d8	Learned
1	1	1	00:12:3f:d8:86:87	Learned
1	8	1	00:13:64:17:14:3a	Learned
1	10	1	00:40:05:52:e1:e7	Learned
			Total = 144 Perm = 0	

Fig. 5.18   Content of the switch database

```
Location        Address        VLAN ID Permanent
Port 3       00-06-5b-73-c0-d8     1      no
Port 1       00-12-3f-d8-86-87     1      no
Port 8       00-13-64-17-14-3a     1      no
Port 10      00-40-05-52-e1-e7     1      no

Select menu option (bridge/port/address):
```

A basic switch management task is to disable a port, because it is either unused or misused. This is done as follows with port 10:

```
-----------------------------------------------------------
Select menu option: ethernet

Menu options: -----------------------------------------------
 autoNegotiation    - Enable/Disable auto-negotiation
 flowControl        - Enable/disable flow control
 portMode           - Set the port speed and duplex mode
 portState          - Enable/Disable ports
 statistics         - Statistics for a port
 summary            - Display port stats summary

Type "q" to return to the previous menu or ? for help.
-----------------------------------------------------------
Select menu option (ethernet): portState
Select Ethernet port (1-24): 10
Enter new value (enable,disable) [enable]: disable

Select menu option (ethernet):
```

When port 10 is disabled, the Linux box is not able to contact host C anymore, as shown by the following arping command execution:

```
[root@linuxbox ~]# arping -b -c 1 192.168.20.50
ARPING 192.168.20.50 from 192.168.20.2 eth0
Sent 1 probes (1 broadcast(s))
Received 0 response(s)
[root@linuxbox ~]#
```

Other management options include manually adding or removing MAC addresses to the switch database, tuning port parameters, enabling traffic monitoring, administering STP and configuring VLANs.

5.10.2   VLAN configuration

By default, all switch ports are part of a predefined virtual LAN, in our example with VID = 1. The following lines show the typical procedure used to create another VLAN with VID = 2:

```
------------------------------------------------------------
Select menu option (bridge): vlan

Menu options: -----------------------------------------------
 addPort            - Add a port to a VLAN
 create             - Create a VLAN
 delete             - Delete a VLAN
 detail             - Display detail information
 modify             - Modify a VLAN
 removePort         - Remove a port from a VLAN
 summary            - Display summary information

Type "q" to return to the previous menu or ? for help.
------------------------------------------------------------
Select menu option (bridge/vlan): create
Enter VLAN ID (2-4094) [2]: 2
Enter Local ID (2-16) [2]: 2
Enter VLAN Name [VLAN 2]: VLAN2
Creating VLAN.

Select menu option (bridge/vlan): summary
Select VLAN ID (1-4094, all) [1]: all

VLAN ID          Local ID          Name
  1                 1               Default VLAN
  2                 2               VLAN2

Select menu option (bridge/vlan):
```

Then it is possible to move ports from the default VLAN to the new one, ports 3 and 8 in our case:

```
Menu options: --------------------------------------------
 addPort              - Add a port to a VLAN
 create               - Create a VLAN
 delete               - Delete a VLAN
 detail               - Display detail information
 modify               - Modify a VLAN
 removePort           - Remove a port from a VLAN
 summary              - Display summary information

Type "q" to return to the previous menu or ? for help.
----------------------------------------------------------
Select menu option (bridge/vlan): addPort
Select VLAN ID (1-4094) [1]: 2
Select Ethernet port (1-24, all): 3
Enter tag type (none, 802.1Q) [802.1Q]: none

Select menu option (bridge/vlan): addPort
Select VLAN ID (1-4094) [1]: 2
Select Ethernet port (1-24, all): 8
Enter tag type (none, 802.1Q) [802.1Q]: none

Select menu option (bridge/vlan):
```

When moving ports to VLAN2, the system asks the manager to specify whether or not 802.1Q tagging must be enabled. In our case, the tagging function is not used since the two ports are connected to ordinary Ethernet stations. The switch database now looks like this:

```
----------------------------------------------------------
Select menu option (bridge/port/address): list
Select bridge ports (1-24, all): all

Location        Address              VLAN ID Permanent
Port 1          00-12-3f-d8-86-87       1       no
Port 10         00-40-05-52-e1-e7       1       no
Port 3          00-06-5b-73-c0-d8       2       no
Port 8          00-13-64-17-14-3a       2       no

Select menu option (bridge/port/address):
```

Launching the arping command on the Linux box toward the other three hosts results in a positive ARP reply from host C only, since hosts A and B are now on a different VLAN:

```
[root@linuxbox ~]# arping -b -c 1 192.168.20.1
ARPING 192.168.20.1 from 192.168.20.2 eth0
Sent 1 probes (1 broadcast(s))
Received 0 response(s)
[root@linuxbox ~]# arping -b -c 1 192.168.20.50
ARPING 192.168.20.50 from 192.168.20.2 eth0
```

```
Unicast reply from 192.168.20.50 [00:40:05:52:E1:E7]   0.734ms
Sent 1 probes (1 broadcast(s))
Received 1 response(s)
[root@linuxbox ~]# arping -b -c 1 192.168.20.254
ARPING 192.168.20.254 from 192.168.20.2 eth0
Sent 1 probes (1 broadcast(s))
Received 0 response(s)
[root@linuxbox ~]#
```

5.10.3   Inter-VLAN communication

As already discussed, inter-VLAN communication requires the use of layer-3 equipment. In our example, such a device interconnecting VLAN1 and VLAN2 can be realized with a Linux box equipped with two NICs. However, this solution is not the most efficient one, since two ports of the switch have to be used. The alternative is to enable the 802.1Q support in the Linux kernel and to use a single connection to the switch that, exploiting the 802.1Q tagging, carries frames for both VLANs. When the Linux kernel supports 802.1Q, the `vconfig` command becomes available to the administrator. This command allows the definition of multiple virtual network interfaces on top of a single physical interface and the desired VID to be set for each one of them. In the following example, VLANs with VID = 1 and 2 are added on top of `eth2`. This results in two new virtual interfaces, named `eth2.1` and `eth2.2`, which are assigned two IP addresses according to the topology shown in Fig. 5.19.

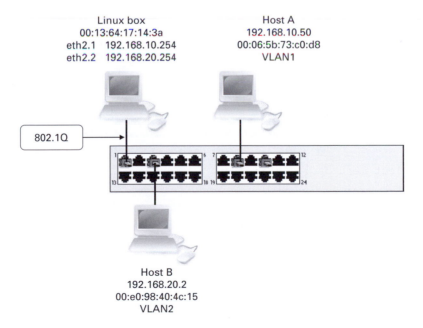

Fig. 5.19   Inter-VLAN set-up

```
[root@linuxbox ~]# vconfig add eth2 1
Added VLAN with VID == 1 to IF -:eth2:-
[root@linuxbox ~]# vconfig add eth2 2
Added VLAN with VID == 2 to IF -:eth2:-
[root@linuxbox ~]# ifconfig eth2.1 192.168.10.254
[root@linuxbox ~]# ifconfig eth2.2 192.168.20.254
[root@linuxbox ~]#
```

The switch must be configured to apply 802.1Q tagging on port 1:

```
------------------------------------------------------------
Select menu option (bridge/vlan): addPort
Select VLAN ID (1-4094) [1]: 1
Select Ethernet port (1-24, all): 1
Enter tag type (none, 802.1Q) [802.1Q]: 802.1Q

Select menu option (bridge/vlan): addPort
Select VLAN ID (1-4094) [1]: 2
Select Ethernet port (1-24, all): 1
Enter tag type (none, 802.1Q) [802.1Q]: 802.1Q

Select menu option (bridge/vlan):
```

Capturing a simple ping session between the Linux box and host B on the Linux tagged interface eth2 shows the actual 802.1Q frame format, as shown in Fig. 5.20. Enabling IP forwarding on the Linux box allows the two VLANs to communicate:

```
[root@linuxbox ~]# route -n
Kernel IP routing table
Destination     Gateway     Genmask         Iface
192.168.20.0    0.0.0.0     255.255.255.0   eth2.2
192.168.10.0    0.0.0.0     255.255.255.0   eth2.1
169.254.0.0     0.0.0.0     255.255.0.0     eth2
```

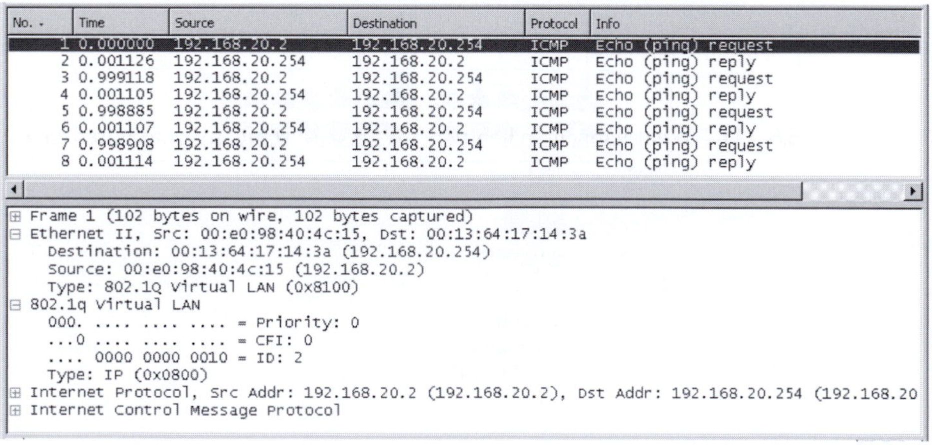

Fig. 5.20   Capture of IEEE 802.1Q tagged frames

```
[root@linuxbox ~]# sysctl -w net.ipv4.ip_forward=1
net.ipv4.ip_forward = 1
[root@linuxbox ~]#
```

In fact, a traceroute command executed on host A toward host B shows the correct forwarding through the Linux box:

```
[user@hostA ~]$ traceroute -n 192.168.20.2
traceroute to 192.168.20.2, 30 hops max, 40 byte packets
 1  192.168.10.254   0.419 ms    0.805 ms    0.906 ms
 2  192.168.20.2   0.157 ms    0.114 ms    0.117 ms
[user@hostA ~]$
```

5.11   Exercises

5.1   Q – What is the spanning tree algorithm?

A – The spanning tree algorithm aims at determining a tree topology removing closed loops by selectively disabling output ports of layer-2 devices, such as bridges. At the end, any couple of nodes of the topology is connected through one single path only.

5.2   Q – What is the difference between a bridge and a switch?

A – They are both layer-2 devices. A switch can be seen as a multiport bridge designed and developed to avoid bandwidth bottlenecks in a LAN. A switch can be considered as a powerful bridge, equipped with processors and memories capable of supporting and processing multiple parallel data flows.

5.3   Q – What is the relation between the broadcast domain and the collision domains of an isolated switch?

A – A broadcast domain can contain several collision domains, but not vice versa.

5.4   Q – Refer to Fig. 5.6 again and imagine that you wish to plug an additional laptop to port 7 of the switch. What ports will receive the frames that the laptop transmits?

A – If the VLAN configuration of the switch is not modified, only the unused ports 6 and 8 will receive the frames: poor thing, nobody will listen to its burbling!

5.5   Q – What happens when two VLANs are statically configured in a switch where a hub, rather than a single station, is connected to one of the switch ports? Refer to the situation in Fig. 5.15, where a hub is connected to the first port of the switch.

A – When the static configuration is adopted, the entire LAN behind the port or hub belongs to the VLAN the port is a member of. Hence, stations A1 through A3 belong to VLAN1, as well as stations B and C.

5.6   Q – Go back to the architecture illustrated in Fig. 5.7. Could you think of an alternative approach to allow intra-VLAN communication? Comment on its drawbacks.

A – The single connection between the two switches, shared by the two VLANs, could be replaced by two distinct connections, one exclusively dedicated to VLAN1, the other

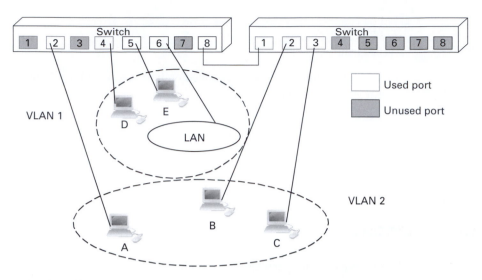

Fig. 5.21 A simple architecture where only one VLAN spans over the two switches

to VLAN2. However, a total of four ports would be requested, as opposed to the two ports that the shared connection in Fig. 5.7 employs; scalability is definitely not the keyword to associate with this trivial approach. Incidentally, this solution requires no frame tagging.

5.7 Q – Focus on the bridged LAN whose physical layer topology is the one reported in Fig. 5.21. Does the switch on the right need to know the VLAN1 database? Is tagging required for the frames that the two switches exchange?

A – The switch on the right does not have to know the VLAN1 database! As for tagging, even if only VLAN2 frames traverse the shared connection, both switches will tag the frames forwarded onto this common trunk.

5.8 Q – What remedies can be adopted if a switch does not have enough memory to run the databases that both the static and dynamic configuration of VLANs requires?

A – If the network administrator runs into this problem, one solution could be to increase the RAM that the switch is equipped with.

Acknowledgments Maria Luisa Merani wishes to thank Michele Borri for the fruitful discussions and cooperation during the first writing of the sections regarding VLANs.

6 Routers

In the previous chapters, a few network devices have been presented. Repeaters are layer-1 devices that perform functions related to physical signal processing. Bridges are layer-2 devices that are able to detect frames and interpret their fields, such as MAC addresses. A switch is a multiport bridge, still a layer-2 device, but with high frame-processing performance.

Even if LAN interconnections in wide areas may be done by using bridges, and until the early 1990s this was quite common, the best way for managing wide area connections is through layer-3 devices, usually called *routers*. Routers perform all functions from layer-1 up to layer-3: they can process packets and are the main devices for network layer management. Indeed, routers represent the most important intermediate systems used to interconnect heterogeneous networks on a geographical scale, building a consistent and cooperative complex system commonly known as an internet.

This chapter illustrates the main concepts behind IP-based routers, discussing the most important functions performed and describing the most common architectures available. Some details of the IP packet processing and forwarding operations are provided, followed by the description of more advanced network layer functionalities, such as packet filtering and address translation. The chapter is concluded by a few practical configuration examples.

6.1 What is a router?

A possible way to start presenting routers is to try to compare them with bridges [65]. A bridge physically connects different LANs but, actually, they can be logically considered the same network, with the same network rules and the same network protocol. A router, on the other hand, completely separates LANs and networks, both physically and logically. While a bridge is a plug-and-play device, which basically works as soon as it is switched on, a router is a "specialized computer" equipped with an operating system that must be properly programmed to do its job. Routers compute and manage data paths through the network, ensuring path redundancy as well. In addition, they can be used as filters to protect the network against unauthorized accesses and everything else that can damage the network.

Considering again the differences between a bridge and a router, a bridge can only perform single hop packet forwarding, whereas a router can perform multihop packet forwarding. Bridges do not implement functions such as hop count, fragmenting and reassembling and congestion notification, which are performed by routers. As far as packet forwarding is concerned, bridges do not reorder packets, while routers do.

As regards addresses, a bridge can just use layer-2 MAC addresses, which do not have any topological meaning and are preassigned by manufacturers. In addition, all broadcast frames are sent to all output ports, including geographical ones. This may create broadcast storms with a lot of control frames occupying bandwidth with no use. Routers work on network and internetwork addresses, such as IP addresses, which are managed within LANs by network managers.

Routers can compute and determine the best source–destination path following a given metric, they are fast at reconfiguring data paths after events such as link or node failures and they do not fix upper bounds on the number of hosts. If required by interconnected networks, they can fragment and reassemble packets, they block broadcast storms and they can manage network congestion events. In addition, routers are the main devices for designing architectures for network security.

Most of the routers currently deployed, we could say all, work with the IP protocol because they operate in the Internet. If we take a look at the Internet, its distributed and scalable nature pushes for a fast and out-of-control growth of users, hosts, access lines and applications, as testified for instance by the estimated number of Internet hosts, which in January 2009 reached 625 million units[1]. Therefore, the exponential growth of traffic volumes asks for ever larger core network capacities.

Network providers face, then, a lot of challenges, ranging from very high-speed infrastructures to ever more powerful switch or router architectures, with switching performance capable of sustaining the ever increasing fiber transmission speeds. The evolution of optical fibers for transmission systems has moved the network bottleneck for end-to-end data communications from lines to nodes. *Dense wavelength division multiplexing* (DWDM) systems allow 128 OC-192 (10 Gbit/s) channels in a single fiber, for an overall 1.2 Tbit/s link speed. There are already fibers working at OC-768 (40 Gbit/s).

Thus, it is clearly fundamental to develop solutions and technologies that allow the design and construction of core IP switches and routers operating at least at terabit/s speed [67]. This must be done together with a strict control on latency and packet loss, to guarantee the quality of service levels required by the different user applications.

6.2   Functions and architectures

The IP version 4 is a connectionless network layer protocol that has no means for providing quality of service. The transfer of datagrams from source to destination is done hop-by-hop through network nodes with a "best-effort" approach, which means that routers will do their best to forward packets toward their final destination trying to avoid losses and to reduce delays, but without any performance guarantee.

---

[1] Source: *Internet Systems Consortium*, https://www.isc.org/solutions/survey.

When an IP datagram arrives at one of the input interfaces, the router has first to decide on which of the output interfaces it has to be forwarded by processing its IP header and then it has to transfer the packet according to this decision. These key router operations are implemented by a number of functions that can be classified as either *data path functions* or *control functions*.

Control functions concern router configuration and management and routing information exchanges. A router exchanges information about network topology with other routers by means of routing protocols, such as RIP, OSPF and BGP, and uses it to build the *routing table*. From the routing table it derives a *forwarding table*, which will be used by a forwarding engine to decide how to route packets.

Data path functions are applied on each incoming datagram and consist of the forwarding decision, the switching through the backplane and the output scheduling. When a datagram gets to an input interface, the forwarding engine reads the destination IP address from the header and processes it using the forwarding table. Each routing and forwarding table entry is called a *route* and it represents one of the possible destinations, known by the router, to send packets to. Each entry includes the following information:

- An IP prefix, specifying the destination of the route;
- A 32-bit netmask, specifying the size of the prefix;
- The IP address of the next-hop, i.e., the neighbor to which packets directed to the prefix must be forwarded;
- The output interface to be used to forward the packets directed to the prefix.

The destination IP address taken from the incoming datagram's header is processed by executing a bitwise logical AND operation with the netmask of each table entry. The result is compared with the prefix and, if they match, it means that a suitable route is found and the packet can be forwarded to the corresponding next hop through the specified interface. Fig. 6.1 illustrates an example of a forwarding table where two routes out of three match the destination of the incoming packet.

The logical AND operation is performed starting from the entry with the largest netmask. Therefore, if several routes would match the destination IP address, the most specific one is chosen. This look-up method is called *longest-prefix match* and assumes that precedence is given to longer prefixes. If an entry with prefix and netmask made of

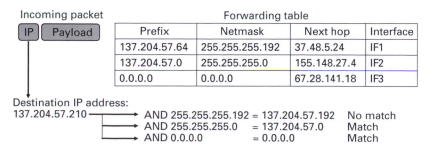

Incoming packet	Forwarding table			
IP Payload	Prefix	Netmask	Next hop	Interface
	137.204.57.64	255.255.255.192	37.48.5.24	IF1
	137.204.57.0	255.255.255.0	155.148.27.4	IF2
	0.0.0.0	0.0.0.0	67.28.141.18	IF3

Destination IP address:
137.204.57.210 ⟶ AND 255.255.255.192 = 137.204.57.192   No match
AND 255.255.255.0   = 137.204.57.0   Match
AND 0.0.0.0   = 0.0.0.0   Match

Fig. 6.1   Forwarding table and address look-up

all zeros is present in the table, it represents a route that matches any possible IP address and, as a consequence, the router knows how to route any possible packet. Since this route is the last one to be processed, it will match all the packets for which no more specific route could be found in the table. This is the reason why such a route is called the *default route*.

Once the output port is determined, a datagram is forwarded to it through the internal switch fabric. Contentions may happen on output ports due to multiple arrivals, so packets have to be temporarily stored inside memories to reduce the packet loss events as much as possible.

A router can also classify packets as a function of their source and destination IP addresses, TCP or UDP port numbers and other upper-layer protocol information. Some datagrams might also be discarded because they do not meet predefined profiles or control checks, or they might be differently managed as a function of their priority level. Other header-processing tasks include time to live (TTL) field update and header checksum computation.

Strictly related to the routing functions described here is the router internal architecture, whose structure mainly depends on the purpose of the router itself, which could interconnect a small set of LANs or be one of the core nodes in a large backbone network.

Entry-level routers are the cheapest routing devices. Forwarding decision and switching functions are executed by a central CPU with a memory placed on a bus architecture (Fig. 6.2) and all the processing is done via software. This means simple architecture and low cost but also low performance, since one CPU has to do all the work and the bus comes to be a bottleneck for data throughput.

To get around performance limits given by only one processor, intermediate-level routers (Fig. 6.3) are enhanced by network cards that are equipped with CPUs and memories employed for packet forwarding. This drastically reduces the central CPU processing load because incoming datagrams are processed in the cards first and then sent to the output network card through the shared bus. Performance in terms of throughput (packets per second) is improved but by increasing the number and speed of input–output interfaces, the bottleneck of the shared bus and central CPU remains.

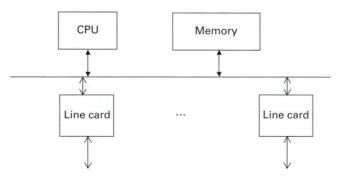

Fig. 6.2   Architecture of entry-level routers

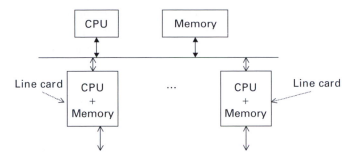

Fig. 6.3 Architecture of intermediate-level routers

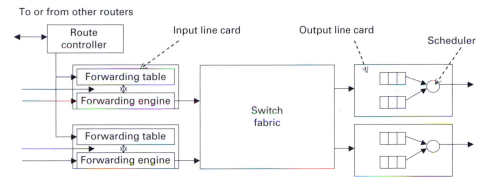

Fig. 6.4 Architecture of high-performance routers

High-performance routers present the most complex architecture (Fig. 6.4). A forwarding engine with its forwarding table is located in each network card and the switching function is done through a dedicated switching system. In essence, a high-performance router is made of a lot of dedicated devices: route controller, forwarding engines, switch fabric and schedulers. Line cards can interact so as to increase the overall throughput. A packet is switched and forwarded to the desired output card and is stored in its output memory to solve possible output contentions. Each output line is managed by a scheduler, which may serve queued packets in accordance to different service policies, also operating with several physical or virtual output queues.

6.3 Table look-up implementation

Routers to be used in today's networks must have the capacity to support the very high speed lines that are in place. Very high speed lines are also crucial for carrying multimedia data, such as the ones typical of peer-to-peer data transfer, video conferencing, and so on.

One of the main design goals of routers is to implement effective and fast techniques for packet forwarding to output ports through the switch fabric. In high-performance routers these techniques are implemented in hardware and forwarding decisions have to be taken by the forwarding engine in a time interval shorter than the shortest datagram duration (wire speed functioning). The internal switch fabric has to give a transit delay as short as possible and with a very low loss probability, thus providing an overall throughput close to 100%.

Routers have to do a lot of tasks, such as datagram header extraction and insertion, updating of the time to live field, datagram classification, and so on. However, the main and most time-consuming task is the IP route look-up, consisting in the search for next hop information in the forwarding table that best fits the IP destination address field of the current datagram.

The forwarding table actually used by a router is structured as a database and it is often referred to as a *forwarding information base* (FIB). The main goal of the design of this database is to minimize the look-up time. In particular, two parameters have to be minimized, the number of memory accesses during the look-up phase and the database size. Reducing the number of accesses is very important because access time is usually high and it represents the bottleneck of look-up procedures. As regards the database size, trying to keep it as small as possible means to be able to put it into a high performance cache memory within the CPU, remarkably reducing access times with respect to traditional random access memories. However, even if the database is too large to be entirely stored in a cache, it is possible to put in it part of the structure, the one which is mostly supposed to be used.

To understand how FIBs are structured, let us consider the representation of a binary tree (Fig. 6.5) spanning the entire space of IP addresses. Since IP addresses are given by 32 bits, the tree height will be 32 and the number of leaves 2^{32}, one for each IP address. Each given prefix entry in the routing table defines a path through the tree that ends in a certain node, so that all the IP addresses, i.e., the leaves, in the sub-tree rooted at this node will have to be routed in the same manner. As a consequence, each entry in the forwarding table defines a set of IP addresses that have the same routing information, i.e., the same next-hop and output interface. If, on the other hand, several table entries meet the same IP address, the longest prefix match rule is applied: given an IP address, the entry which goes most in depth down in the tree, i.e., the closest one to the bottom of the tree, is chosen.

The FIB is stored in a *trie* structure, which is a multipath tree where each node has either zero or more pointers to child nodes. In one-bit trie structures, each node has

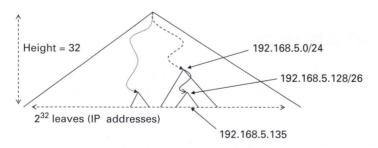

Height = 32

192.168.5.0/24

192.168.5.128/26

2^{32} leaves (IP addresses)

192.168.5.135

Fig. 6.5   Representation of the IP address space as a binary tree

up to two children, corresponding to 0 and 1 respectively. A generic node X at level h represents the set of all paths (i.e., bit sequences or IP addresses) that have the same first h bits. Depending on the $(h + 1)$th bit value, a pointer could go toward a different sub-trie, if it exists, which in turn represents the set of all paths with the same first $h + 1$ bits. Every address look-up begins from the trie root and during the crossing of the trie the node corresponding to the next hop with the longest matching prefix is tracked.

An example of a trie structure is shown in Fig. 6.6. The processing of the address 10110011 leads to node 10 and the longest prefix match returns 10. If address 00110110 has to be processed, 0011 is returned. An underlined node in the figure means that the corresponding prefix has an associated entry in the routing table. A drawback of this mechanism is that in the worst case 32 memory accesses are required.

A typical implementation of routing tables is by means of *Patricia tries*, which are an improvement on the one-bit trie structure. The main idea behind Patricia tries is that

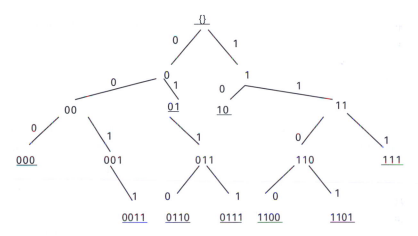

Fig. 6.6   Example of a trie structure

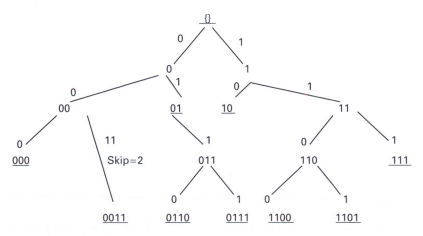

Fig. 6.7   Example of a Patricia trie

a node that has not an associated entry in the table and has just one child node can be removed to shorten the look-up phase. It is, however, necessary to keep track of these removed nodes and this is done by using a number, the skip value, which indicates how many bits are skipped along the path.

An example of a Patricia trie corresponding to the example of Fig. 6.6 is shown in Fig. 6.7. If address 00110110 has to be processed, 0011 is returned again but now with three memory accesses rather than four.

6.4   From routers to middleboxes: firewalls and NATs

The functions previously described represent the normal behavior of IP routers as intermediate devices interconnecting two hosts at the end points of a path across an IP network. However, owing to a number of issues, including network security, efficient use of the IP address space and communication performance, some intermediate devices may show a behavior that deviates from the implementation of standard IP routing functions. To distinguish such non-standard intermediate nodes from traditional IP routers, RFC 3234 [68] adopts the term *middlebox*.

One of the most common forms of middlebox is represented by a *firewall*, a device – or a set of devices – whose main function is to interconnect two or more network segments while inspecting the crossing traffic and allowing only the authorized packets to pass through, according to a suitable set of rules [69]. A firewall may apply its filtering functions at different network levels. A firewall operating at the network and transport layers is typically called a *packet filter*, since its job is to filter packets based on the information included in the IP and TCP or UDP headers. On the other hand, an *application-level gateway* is a firewall able to inspect the entire contents of each packet and to take the filtering decision based on information related to the application-layer protocol used. In line with the general network orientation of this book, in this section we will focus on packet filters only.

A packet filter may be implemented either as a *stateless* or a *stateful* device. Since stateless filters do not keep track of ongoing TCP connections or UDP sessions, each packet is seen as an independent entity and must be treated separately. On the contrary, stateful filters are able to discriminate when a new connection or session is initiated, e.g., by inspecting the SYN flag in TCP headers or by detecting when a packet from an unknown UDP session is seen for the first time. Then they temporarily store the tuple identifying each new connection or session and maintain its state until the connection is explicitly closed – e.g., by inspecting the FIN flag in the TCP header – or a timeout expires since the last packet belonging to the connection or session is detected. The tuple used to identify a session or connection is given by the source IP address and port, the destination IP address and port and the transport protocol adopted. Using this information, each packet crossing the filter is considered as part of a specific session or connection and is, therefore, treated accordingly.

Although the previous considerations assume the use of a transport protocol defining port numbers, such as TCP and UDP, a session state can also be defined when other protocols are encapsulated inside IP packets, granted they provide a way to distinguish between different sessions. For instance, ICMP messages are directly encapsulated in IP packets and port numbers are not used. However, the ICMP header includes identification and sequence number fields that can be used to discriminate different ICMP sessions.

The packet filter behavior is typically defined through a set of rules. Each rule determines to which packets it has to be applied by specifying the value in the relevant IP and transport header fields. Then each rule specifies what action must be taken on the matching packets, such as allowing its transit or discarding it. If no matching rule can be found for a given packet, a default policy is applied that can be defined either as *default allow*, where any unmatching packet is let through, or *default deny*, meaning that any packet whose transit is not explicitly allowed by a rule is discarded.

When a stateless filter is used, there is no memory of the session or connection state and the whole set of rules must be applied to every packet crossing the device. This means that when the transit of a given tuple is authorized by a rule in a given direction, another rule must be explicitly defined to allow the transit of the corresponding tuple in the opposite direction, i.e., for packets where source and destination addresses and port numbers are swapped. On the other hand, when a stateful filter is adopted, packets that initiate a new connection or session are detected and the following packets in both directions are identified as part of an existing connection or session and treated as such. The advantage of the stateful approach is that the set of rules to be specified can be limited to the packets that initiate a new connection or session, since the firewall itself is able to detect whether a packet is part of an existing traffic flow or is invalid.

A typical example of network set-up including a packet filter will be described later in this chapter.

Another very common class of middlebox includes all those intermediate devices that modify packet headers by translating private IP addresses into public ones and vice versa. As already discussed in Section 2.4.1, for security reasons, cost-reduction policies and lack of IPv4 address space, the use of private IP addressing to configure the internal networks of an institution or company, and the consequent adoption of address translation techniques to reach the public Internet, have become very popular and widespread. To provide a standard way of implementing address translation between private and public IP networks, the IETF decided to tackle this problem and released RFC 1631 in the mid 1990s [70], which was later extended by RFCs 2663 [71] and 3022 [72]. In general, a *network address translator* (NAT) is a middlebox capable of modifying the address fields of an IP packet while forwarding it between different IP networks. This definition does not strictly require the use of private IP addresses, although NATs are mainly used to allow hosts with a private IP address to exchange packets with hosts using public addresses.

The most simple type of NAT is called *basic NAT*; it involves IP address translation only and ignores information related to higher-layer protocols. Let us consider the topology shown in Fig. 6.8, where hosts A and B on the internal network use the private prefix 192.168.8.0/24 and are able to reach the rest of the Internet through a basic NAT device, which has been configured with a private address on the internal interface and two public addresses on the external one. When host A sends a packet to host C, it must use the NAT as a next hop. The NAT itself, before transmitting the packet through the external

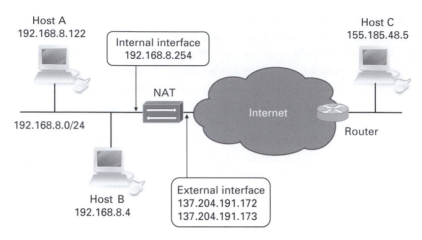

Fig. 6.8   NAT topology

interface, modifies the IP source address field and replaces 192.168.8.122 with one of its public addresses, for instance 137.204.191.172. Then the packet is routed through the Internet and reaches host C, which believes that the packet has been originated by a host with public address 137.204.191.172. Therefore, if C has some data to send back to A, it generates a packet with destination address 137.204.191.172, which is routed back to the external interface of the NAT. Then the NAT forwards the packet to host A, but only after having modified the destination address field and replaced the public address with 192.168.8.122. Both the endpoints are not aware of the address translation, since basic NAT is a completely transparent operation.

When different hosts on the same private network have packets to be sent to the public Internet, the NAT must keep track of the translations performed, to guarantee connectivity. This can be done either statically or dynamically. In the former case, a fixed correspondence between private and public address is defined. For instance, in Fig. 6.8 host A address 192.168.8.122 will always be translated into 137.204.191.172 and host B address 192.168.8.4 into 137.204.191.173. Any packet coming back from the NAT external interface and directed to any of the two public addresses will be forwarded to the corresponding internal host after the private address has been restored. Obviously this solution has the drawback that, to allow any further host on the private network to access the public realm, an additional public address must be assigned to the NAT, so the number of private hosts allowed to reach the public Internet is limited by the number of NAT public addresses.

On the other hand, if a dynamic address translation is adopted, the same public address can be used by different private hosts at different times. For instance, when the traffic exchange between hosts A and C terminates, the public address used for the translation becomes available and can be used to translate flows involving a different private host. With this approach, the number of public addresses assigned to the NAT limits the number of private hosts that are able to access the public realm *simultaneously*. However, this solution requires the NAT device to be able to keep track of any traffic flow forwarded

Table 6.1. Basic NAT translation table

Internal		External
192.168.8.122	$\Longleftrightarrow$	137.204.191.172
192.168.8.4	$\Longleftrightarrow$	137.204.191.173

between the private and the public interfaces, where flows correspond to, e.g., TCP connections and UDP or ICMP sessions.

The basic NAT maintains a translation table with the known correspondences between private and public addresses. For the example of Fig. 6.8 such a table could be as shown in Table 6.1. While in the static case the table entries are fixed and manually configured, in the dynamic case they appear when a new flow is detected and disappear when the corresponding flow is considered expired. In any case, the number of table entries cannot exceed the number of public addresses assigned to the NAT.

A more flexible NAT solution is the *network address and port translation* (NAPT), which involves not only IP address translation, but also modifications to higher-layer protocol information, such as TCP and UDP ports. This approach allows simultaneous traffic flows involving different private hosts to share the same public address, since the NAT is now able to keep track of the translations by looking also at the TCP or UDP port values. Typical NAPT implementations try to avoid modifying port values as much as possible, but they are allowed to do it when this is necessary to discern between different flows.

For instance, assume that the NAT in Fig. 6.8 has been assigned 137.204.191.172 only as a public address. Also let us assume that private host A has already established a TCP connection with public host C, identified by the address:port couples 192.168.8.122:3923 and 155.185.48.5:80. When forwarding such packets, the NAT translates the source address into 137.204.191.172 and, if local TCP port 3923 is not already used by other connections, it leaves the source port value unchanged. Then assume that host B starts a new TCP connection toward the public realm using source port 4109. The NAT translates the private source address 192.168.8.4 again into 137.204.191.172 and, since port 4109 is not used by other connections, it leaves this information unchanged. Finally after a while, host A starts a new connection using source port 4109. This time the NAT must translate both source address and port, since value 4109 is already in use by the connection involving hosts B and C. Therefore, the NAT translates the source address:port couple into 137.204.191.172:5196, where the 5196 source port value is not used by other connections and is chosen according to some internal criteria. The NAT translation table related to the TCP connections of the previous example is shown in Table 6.2.

The number of simultaneous connections originated from private hosts and directed to the public realm is now limited by the total number of TCP or UDP ports available on the NAT external interface, i.e., 2^{16} for each public address assigned to the NAT. Therefore, with a single public address it is possible to provide connectivity to a quite large private network. An NAPT implementation using a single public IP address is also known as *port address translation* (PAT) or *IP masquerading*.

Table 6.2. NAPT translation table

Internal			External	
Address	Port		Address	Port
192.168.8.122	3923	⟺	137.204.191.172	3923
192.168.8.4	4109	⟺	137.204.191.172	4109
192.168.8.122	4109	⟺	137.204.191.172	5196

Table 6.3. Port forwarding translation table

External			Internal	
Address	Port		Address	Port
137.204.191.172	80	⟺	192.168.8.122	80
137.204.191.172	25	⟺	192.168.8.4	25

The example discussed here describes the case of an outbound NAT, where the main purpose is to allow hosts on a private network to contact hosts on the public Internet, with traffic flows originated from the private hosts. In this scenario, inbound connections initiated from public hosts toward the private network are generally not allowed, keeping the private hosts secure from external attacks. However, there are some situations where it is required that specific services available on private hosts are accessible by external client applications, so an inbound NAT must be performed. This is straightforward when basic NAT with static translations is used. In the case of dynamic translations or NAPT, this behavior must be specifically configured inside the NAT, telling the device that any packet directed to a given address:port couple on the external interface must be forwarded to a specific private host after the suitable translation. This technique is known as *port forwarding*.

For instance, in Fig. 6.8, if public web and email servers are hosted by A and B respectively, they must be accessible from the outside. Therefore, the NAT must be configured with a port forwarding table such as the one shown in Table 6.3.

Address translation techniques are supposed to be transparent to the endpoints involved in the communication, since application level protocols should not be concerned about network and transport layer information. However, there are some applications that break this rule and include addresses or port values in the payload. For instance, file transfer protocol (FTP) [73] control sessions include the negotiation of the ports to be used for the data transfer connection: an address and port change caused by a NAT device may compromise the correct protocol behavior. Another example is given by the IPsec standard [74], used to set up network-layer virtual private networks: a modification in the address field by a NAT may cause the failure of the IPsec packet integrity check. Network address translators may also cause problems to peer-to-peer protocols and to voice over IP and multimedia traffic transported by UDP. For further details about

NAT-related issues see RFC 3027 [75]. Solutions to such problems include: making the NAT aware of application protocol behavior by implementing application-level gateways, adopting suitable NAT-traversal strategies and enforcing NAT-discovery protocols, such as STUN [76]. Recently, RFC 4787 [77] defined the recommended NAT behavior that would allow UDP-based application protocols to work in the presence of address translation.

For the latest developments on the use of middleboxes in the Internet, see [78].

6.5   Practice: routing and forwarding table

One of the most useful networking tools offered by the Linux kernel is the IP packet forwarding function, which allows a multi-homed Linux box to act as a router by forwarding packets incoming on a network interface to a different outgoing interface, based on the packet's destination address. The choice of the output interface is made using a look-up operation on the kernel routing table that includes information on how to route packets directed to known destinations. With the capability of packet forwarding and the use of specific software implementing the most common routing protocols, as will be discussed in the next chapter, a Linux box can be used as a fully functional IP router. Of course it must be pointed out that a commercial router is typically a piece of equipment specifically engineered to perform routing tasks, e.g., by means of specialized hardware, while a Linux-based router can be run on any general-purpose PC and, therefore, offers much lower performance, in particular when processing large routing tables and using interfaces with very high data rates. However, from the functional point of view, a Linux box can be used to perform interesting experiments on IP routing.

The very first thing to do to make a Linux box act as a router is to enable the packet forwarding function inside the kernel with the following command:

```
[root@linuxbox ~]# sysctl -w net.ipv4.ip_forward=1
net.ipv4.ip_forward = 1
```

The `net.ipv4.ip_forward` flag is one of several kernel configuration flags that can be used to modify the behavior of the TCP/IP stack at runtime [79]. The possibility of disabling packet forwarding is a nice security feature to be adopted whenever a multi-homed box is supposed to act as a simple host connected to multiple network segments, in order to avoid any possible misuse, e.g., to bypass a packet-filtering firewall.

Consider the network topology shown in Fig. 6.9, where the three routers are Linux-based. Assume that the host and router interfaces have already been configured with the required IP parameters using the `ifconfig` command, as previously explained in Section 3.5. Assume also that hosts 1 and 2 have been configured to use routers A and B as default gateways, respectively, and that IP forwarding has been enabled on each router. The topology considered here is so simple that it is sufficient to use static routes to configure the routers properly.

Now, since router C is the only device that connects our network to the rest of the Internet, it is reasonable to configure it as the default gateway for routers A and B using

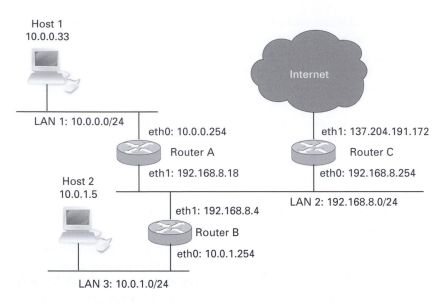

Fig. 6.9   Network set-up with three LANs and three routers

the `route` command on both machines:

```
[root@RouterA ~]# route add default gw 192.168.8.254 dev eth1

[root@RouterB ~]# route add -net 0.0.0.0/0 gw 192.168.8.254 dev eth1
```

These commands are two different ways to specify the default gateway, since the default route corresponds to a destination address and a netmask made of all zeros. At this point, router A knows how to route packets destined to every possible destination. In fact, it knows that packets directed to addresses belonging to LANs 1 and 2 must be directly delivered by simply performing an ARP request toward the final destination, while any other packet must be forwarded to router C for indirect delivery. However, this configuration is incomplete since router A assumes a wrong direction for packets destined to LAN 3, which should be forwarded to router B instead of C. Therefore, we need to let router A know about this by adding a static entry to its routing table, specifying that network 10.0.1.0/24 is reachable via router B using interface `eth1`, as follows:

```
[root@RouterA ~]# route add -net 10.0.1.0/24 gw 192.168.8.4 dev eth1
```

At this point the kernel routing table of router A appears as

```
[root@RouterA ~]# route -n
Kernel IP routing table
Destination      Gateway          Genmask          Flags Metric Iface
10.0.0.0         0.0.0.0          255.255.255.0    U     0      eth0
10.0.1.0         192.168.8.4      255.255.255.0    UG    0      eth1
192.168.8.0      0.0.0.0          255.255.255.0    U     0      eth1
0.0.0.0          192.168.8.254    0.0.0.0          UG    0      eth1
```

Similarly, the static route to be added to router B and the resulting routing table are

```
[root@RouterB ~]# route add -net 10.0.0.0/24 gw 192.168.8.18
[root@RouterB ~]# route -n
Kernel IP routing table
Destination     Gateway         Genmask         Flags Metric Iface
10.0.0.0        192.168.8.18    255.255.255.0   UG    0      eth1
10.0.1.0        0.0.0.0         255.255.255.0   U     0      eth0
192.168.8.0     0.0.0.0         255.255.255.0   U     0      eth1
0.0.0.0         192.168.8.254   0.0.0.0         UG    0      eth1
```

It is worth noticing that in both cases, when adding the static route, the IP address of the router to be used as a next hop is the one assigned to the interface connected to LAN 2, i.e., the address belonging to network 192.168.8.0/24 to which both routers are directly attached. This is mandatory, since every route related to a destination reachable through an indirect delivery must specify a next hop's IP address that is reachable by means of a direct delivery.

To test the correctness of the router configurations, a traceroute command can be executed on host 1 toward host 2 showing the path followed by packets directed to IP address 10.0.1.5 as well as by the replies coming from host 2 and directed to 10.0.0.33:

```
[root@Host1 ~]# traceroute 10.0.1.5
traceroute to 10.0.1.5 (10.0.1.5), 30 hops max, 40 byte packets
 1   10.0.0.254 (10.0.0.254)   0.105 ms    0.067 ms    0.067 ms
 2   192.168.8.4 (192.168.8.4)   2.482 ms    2.301 ms    0.152 ms
 3   10.0.1.5 (10.0.1.5)   0.279 ms    0.220 ms    0.216 ms
```

To have a complete routing configuration on the whole network and assuming that 137.204.191.254 is the next hop toward the Internet, the following static routes must be added to router C's table,

```
[root@RouterC ~]# route add -net 10.0.0.0/24 gw 192.168.8.18
[root@RouterC ~]# route add -net 10.0.1.0/24 gw 192.168.8.4
[root@RouterC ~]# route add default gw 137.204.191.254
```

which will then look like

```
[root@RouterC ~]# route -n
Kernel IP routing table
Destination     Gateway         Genmask         Flags Metric Iface
137.204.191.0   0.0.0.0         255.255.255.0   U     0      eth1
10.0.0.0        192.168.8.18    255.255.255.0   UG    0      eth0
10.0.1.0        192.168.8.4     255.255.255.0   UG    0      eth0
192.168.8.0     0.0.0.0         255.255.255.0   U     0      eth0
0.0.0.0         137.204.191.254 0.0.0.0         UG    0      eth1
```

An alternative that has been recently introduced in Linux kernels is given by the `ip` and `tc` commands [79], which provide the same functionalities as the `ifconfig` and `route` commands, plus a number of additional advanced networking features, such as tunneling, multiple routing table definition, policy routing, packet classification and traffic conditioning. The use of such commands to show router C's kernel routing table is

```
[root@RouterC ~]# ip route show
137.204.191.0/24 dev eth1 proto kernel scope link
10.0.0.0/24 via 192.168.8.18 dev eth0
10.0.1.0/24 via 192.168.8.4 dev eth0
192.168.8.0/24 dev eth0 proto kernel scope link
default via 137.204.191.254 dev eth1
```

A comparison of this output with the output of the `route` command is straightforward. The interesting feature of the `ip route` command is that it is possible to show the so-called *cache routing table*, which is the Linux implementation of the forwarding table. In fact, this cache table stores an entry for each packet that has been successfully forwarded by the kernel after the first look-up on the routing table. Following packets matching the same entry are quickly forwarded based on the cached information and no routing table look-up is required. As an example, when host 1 sends a packet to the external address 137.204.57.26 and gets a reply, the cache routing table on router C looks like

```
[root@RouterC ~]# ip route show cache
67.171.78.2 from 137.204.191.172 via 137.204.191.254 dev eth1
    cache  mtu 1500 advmss 1460 hoplimit 64
local 137.204.191.172 from 67.171.78.2 dev lo
    cache <local>  iif eth1
192.168.8.18 from 192.168.8.254 dev eth0
    cache  mtu 1500 advmss 1460 hoplimit 64
137.204.57.26 from 10.0.0.33 via 137.204.191.254 dev eth1
    cache  mtu 1500 advmss 1460 hoplimit 64 iif eth0
local 192.168.8.254 from 192.168.8.18 dev lo
    cache <local,src-direct>  iif eth0
10.0.0.33 from 137.204.57.26 via 192.168.8.18 dev eth0
    cache  mtu 1500 advmss 1460 hoplimit 64 iif eth1
```

It can be noticed that entries 4 and 6 are related to the forwarding of packets between 10.0.0.33 and 137.204.57.26 and will be used to forward the following packets with the same source and destination addresses.

A few considerations should be added to conclude this section. First, the reader may have noticed that the routing tables shown above include a column named `metric` that specifies the cost of each entry: this allows the inclusion of multiple routes to the same destination using different costs, for instance because the routing protocols adopted found multiple paths toward the same network. The use of the metric parameter ensures that the kernel by default will always choose the route with minimum cost.

Another consideration is related to the IP addresses used in the example of Fig. 6.9: to allow the internal hosts and routers, which are using private addresses, to communicate with recipients in the public Internet, router C should also act as a NAT and avoid routing packets with private source addresses on interface `eth1`. This problem was simply neglected here, since this section focused on the forwarding function only.

6.6   Practice: firewalls and packet filtering

The typical set-up of an IP network including a firewall is depicted in the example of Fig. 6.10. The figure shows the network premises of a small company, including an

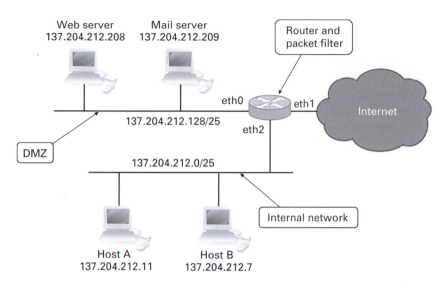

Fig. 6.10   Typical network firewall set-up

internal LAN and a number of servers publicly accessible, like web and email servers. The company's IP network 137.204.212.0/24 has been split into two subnets, as indicated in the figure. The company's policies require two different levels of security. The hosts on the internal network must be protected by any unauthorized access from the Internet, which is considered an untrusted domain, since any kind of attack can potentially come from there. On the other hand, the servers on the so-called *demilitarized zone* (DMZ) must be accessible from the outside, although it is good practice to limit the inbound traffic to the services actually provided by the company to the public. Therefore, the firewall must be configured by specifying a set of rules such that:

- Every connection or session initiated from the outside and directed to the DMZ must be allowed if the destination IP address and port number correspond to a publicly accessible server;
- Every connection or session initiated from the DMZ and directed to the Internet must be allowed;
- Every connection or session initiated from the internal network and directed to the DMZ or to the Internet must be allowed;
- Everything else must be blocked.

A Linux box can be configured to act as a packet filter, using both a stateless and stateful approach. The tool that provides such functionality is the `iptables` command [80], which was integrated in the Linux kernel starting from version 2.4. The `iptables` command is a general purpose tool that allows a user with root privileges to intercept packets and modify most of the IP and transport protocol header fields, thus implementing different kinds of middleboxes. When packets traverse a Linux kernel with `iptables` enabled, they are subject to different processing modes represented by so-called *tables*, each implementing the functionalities of a specific middlebox. Each table is made of a few sets of rules, called *chains*, that specify what action must be taken on packets based on the content of their IP and transport headers. The rules included in a given chain are applied in the order they appear. This is something to take into account when adding rules to a chain, since once a packet matches a rule, its fate is determined by that rule and the following rules are not parsed. For instance, if a rule states that every packet carrying a TCP segment directed to port N must be dropped and if such a rule precedes another one stating that all TCP segments directed to a specific IP address $a.b.c.d$ must be accepted, then segments directed to port N of address $a.b.c.d$ will never be allowed, since they match the first rule.

The table implementing a packet filter is called `filter` and is made of three predefined chains:

- `INPUT`, including the filtering rules to be applied to incoming packets destined to the applications running on the Linux box itself;
- `OUTPUT`, including the filtering rules to be applied to outgoing packets generated by the applications running on the Linux box itself;
- `FORWARD`, including the filtering rules to be applied to packets that are forwarded by the Linux box from one interface to another, not including the loopback.

The basic structure of the filter table used by `iptables` is shown in Fig. 6.11. Other chains can be defined by the user to realize more complex schemes.

Let us see how it is possible to implement the packet filter shown in Fig. 6.10 on a Linux box using `iptables`, after the interfaces have been properly configured and the packet forwarding has been enabled in the kernel. Since we are considering packets that are forwarded between different interfaces, we need to add our rules to the `FORWARD` chain of the `filter` table. The first thing to do is to decide which default policy to use. Since in our case the security requirements are specified in the form of allowing a few types of traffic and blocking everything else, it is reasonable to adopt a default deny policy for the `FORWARD` chain as follows:

```
iptables -P FORWARD DROP
```

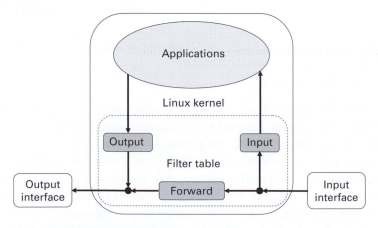

Fig. 6.11   The `filter` table in the Linux kernel

The next step is to enable the transit of authorized packets, for instance those belonging to connections coming from the outside and directed to the web server on the DMZ. The well-known port used by web servers is TCP port 80, so the following command should be executed:

```
iptables -A FORWARD -i eth1 -o eth0 -d 137.204.212.208
        -p tcp --dport 80 -m state --state NEW -j ACCEPT
```

The previous command appends (-A) a new rule to the FORWARD chain stating that every packet incoming on the external interface (-i eth1), outgoing through the DMZ interface (-o eth0), directed to the webserver (-d 137.204.212.208), carrying a TCP segment with destination port 80 (-p tcp --dport 80) and belonging to a connection whose state is unknown to the filter (-m state --state NEW) must be accepted (-j ACCEPT). A similar command must be given again for the web server, assuming that also HTTPS connections (TCP port 443) should be allowed, and another one for the mail server (TCP port 25):

```
iptables -A FORWARD -i eth1 -o eth0 -d 137.204.212.208
        -p tcp --dport 443 -m state --state NEW -j ACCEPT

iptables -A FORWARD -i eth1 -o eth0 -d 137.204.212.209
        -p tcp --dport 25 -m state --state NEW -j ACCEPT
```

Then all connections or sessions initiated from the DMZ or the internal network and directed to the Internet must be allowed:

```
iptables -A FORWARD -i eth0 -o eth1 -s 137.204.212.128/25
        -m state --state NEW -j ACCEPT

iptables -A FORWARD -i eth2 -o eth1 -s 137.204.212.0/25
        -m state --state NEW -j ACCEPT
```

as well as those from the internal network to the DMZ:

```
iptables -A FORWARD -i eth2 -o eth0 -s 137.204.212.0/25
        -d 137.204.212.128/25 -m state --state NEW -j ACCEPT
```

Finally, we need to tell the Linux box to act as a stateful firewall and let go all packets belonging to connections and sessions already authorized:

```
iptables -I FORWARD 1 -m state --state ESTABLISHED -j ACCEPT
```

The previous command inserts the rule on position 1, i.e., on top of the chain (-I FORWARD 1), so every packet that matches that rule is accepted immediately and does not need to be compared with other rules, thus saving some processing time. The resulting set of filtering rules included in the FORWARD chain, shown in the order as they are applied, is as follows:

```
[root@Router ~]# iptables -nv -L FORWARD
Chain FORWARD (policy DROP 1911 packets, 217K bytes)
target  prot in   out   source            destination
ACCEPT  all  *    *     0.0.0.0/0         0.0.0.0/0        state ESTABLISHED
```

```
ACCEPT  tcp  eth1 eth0  0.0.0.0/0           137.204.212.208            tcp dpt:80 state NEW
ACCEPT  tcp  eth1 eth0  0.0.0.0/0           137.204.212.208            tcp dpt:443 state NEW
ACCEPT  tcp  eth1 eth0  0.0.0.0/0           137.204.212.209            tcp dpt:25 state NEW
ACCEPT  all  eth0 eth1  137.204.212.128/25  0.0.0.0/0                  state NEW
ACCEPT  all  eth2 eth1  137.204.212.0/25    0.0.0.0/0                  state NEW
ACCEPT  all  eth2 eth0  137.204.212.0/25    137.204.212.128/25         state NEW
```

An asterisk in the input (`in`) or output (`out`) column means that the rule is applied to packets incoming on any input or outgoing to any output interface respectively. Similarly, a 0.0.0.0/0 prefix in the `source` or `destination` column means any possible IP address, i.e., the packet is not filtered by the rule based on the corresponding IP address field. The last column shows additional filtering parameters such as the connection state and the TCP port numbers.

At this point, the firewall configuration is complete and can finally be tested. For this purpose the `nmap` command proves to be useful. The `nmap` command is a very popular and flexible network scanning tool that allows one to know, among other things, which TCP ports are in the listen state on a remote host by initiating a series of connection attempts toward the desired port range. For instance, launching a TCP port scan from host A on the internal network toward the web server on the DMZ, limiting the port range to (20, 450), gives the following result:

```
[root@HostA ~]# nmap -p 20-450 137.204.212.208

Starting Nmap 4.11 at 2008-06-17 17:56 CEST
Interesting ports on 137.204.212.208:
Not shown: 427 closed ports
PORT     STATE SERVICE
22/tcp   open  ssh
80/tcp   open  http
111/tcp  open  rpcbind
443/tcp  open  https

Nmap finished: 1 IP address (1 host up) scanned in 1.434 seconds
```

Four ports from the scanned range are reported in the open state, including those used by the HTTP and HTTPS services. The remaining 427 are reported as closed, since TCP resets have been received in reply to the connection attempts to those ports. This result is consistent with the firewall configuration, since all traffic originated from the internal network and directed to the DMZ is allowed by the seventh rule in the FORWARD chain.

When the same scan is performed from a host located somewhere in the external network, the result is different:

```
[root@ExtHost ~]# nmap -p 20-450 137.204.212.208

Starting Nmap 4.11 at 2008-06-17 17:59 CEST
Note: Host seems down.
If it is really up, but blocking our ping probes, try -P0
Nmap finished: 1 IP address (0 hosts up) scanned in 2.070 seconds
```

The command output notifies the user that the target host is not replying to the preliminary ping attempts made by nmap. The host is, therefore, considered unreachable and the port scan is not performed. This was actually expected, since ICMP packets are not allowed through the firewall. However, nmap suggests using an additional command-line option (-P0) that prevents the scanner from trying the preliminary ping probes toward the target host and forces the port scan execution.

```
[root@ExtHost ~]# nmap -P0 -p 20-450 137.204.212.208

Starting Nmap 4.11 at 2008-06-17 17:59 CEST
Interesting ports on 137.204.212.208:
Not shown: 429 filtered ports
PORT     STATE SERVICE
80/tcp   open  http
443/tcp  open  https

Nmap finished: 1 IP address (1 host up) scanned in 7.891 seconds
```

Comparing the result of the last scan with the first one, it is apparent that the firewall is doing its job, since only the ports corresponding to the allowed services (HTTP and HTTPS) are shown in the open state, while the other ones are reported as filtered and cannot be contacted from the outside.

Capture on the external interface (eth1)

No. ▲	Time	Source	Destination	Protocol	Info
1	0.000000	137.204.57.222	137.204.212.208	TCP	43637 > http [SYN] Seq=0 Ack=0 Win=5840
2	0.002407	137.204.212.208	137.204.57.222	TCP	http > 43637 [SYN, ACK] Seq=0 Ack=1 Win
3	0.002499	137.204.57.222	137.204.212.208	TCP	43637 > http [ACK] Seq=1 Ack=1 Win=5888
4	0.002868	137.204.57.222	137.204.212.208	HTTP	GET / HTTP/1.1
5	0.002979	137.204.212.208	137.204.57.222	TCP	http > 43637 [ACK] Seq=1 Ack=178 Win=69
6	0.020832	137.204.212.208	137.204.57.222	HTTP	HTTP/1.1 403 Forbidden (text/html)
7	0.020957	137.204.212.208	137.204.57.222	HTTP	Continuation or non-HTTP traffic
8	0.021061	137.204.212.208	137.204.57.222	HTTP	Continuation or non-HTTP traffic
9	0.021162	137.204.57.222	137.204.212.208	TCP	43637 > http [ACK] Seq=178 Ack=1449 Win
10	0.021288	137.204.57.222	137.204.212.208	TCP	43637 > http [ACK] Seq=178 Ack=2897 Win
11	0.022732	137.204.57.222	137.204.212.208	TCP	43637 > http [FIN, ACK] Seq=178 Ack=415
12	0.022812	137.204.212.208	137.204.57.222	TCP	http > 43637 [ACK] Seq=4155 Ack=179 Win
13	27.902155	137.204.57.222	137.204.212.208	TCP	39237 > 22 [SYN] Seq=0 Ack=0 Win=747520
14	30.906035	137.204.57.222	137.204.212.208	TCP	39237 > 22 [SYN] Seq=0 Ack=0 Win=747520
15	36.906453	137.204.57.222	137.204.212.208	TCP	39237 > 22 [SYN] Seq=0 Ack=0 Win=747520
16	48.903290	137.204.57.222	137.204.212.208	TCP	39237 > 22 [SYN] Seq=0 Ack=0 Win=747520
17	72.904972	137.204.57.222	137.204.212.208	TCP	39237 > 22 [SYN] Seq=0 Ack=0 Win=747520

Capture on the DMZ interface (eth0)

No. ▲	Time	Source	Destination	Protocol	Info
1	0.000000	137.204.57.222	137.204.212.208	TCP	43637 > http [SYN] Seq=0 Ack=0 Win=5840
2	0.000097	137.204.212.208	137.204.57.222	TCP	http > 43637 [SYN, ACK] Seq=0 Ack=1 Win
3	0.000205	137.204.57.222	137.204.212.208	TCP	43637 > http [ACK] Seq=1 Ack=1 Win=5888
4	0.000574	137.204.57.222	137.204.212.208	HTTP	GET / HTTP/1.1
5	0.000673	137.204.212.208	137.204.57.222	TCP	http > 43637 [ACK] Seq=1 Ack=178 Win=69
6	0.018524	137.204.212.208	137.204.57.222	HTTP	HTTP/1.1 403 Forbidden (text/html)
7	0.018647	137.204.212.208	137.204.57.222	HTTP	Continuation or non-HTTP traffic
8	0.018754	137.204.212.208	137.204.57.222	HTTP	Continuation or non-HTTP traffic
9	0.018868	137.204.57.222	137.204.212.208	TCP	43637 > http [ACK] Seq=178 Ack=1449 Win
10	0.018993	137.204.57.222	137.204.212.208	TCP	43637 > http [ACK] Seq=178 Ack=2897 Win
11	0.020438	137.204.57.222	137.204.212.208	TCP	43637 > http [FIN, ACK] Seq=178 Ack=415
12	0.020506	137.204.212.208	137.204.57.222	TCP	http > 43637 [ACK] Seq=4155 Ack=179 Win

Fig. 6.12 Simultaneous captures on DMZ and external interface allow to asses the correctness of the firewall configuration

The correct filtering behavior of the firewall is also testified by comparing the packet captures, simultaneously performed on both the external and DMZ interfaces of the router, as shown in Fig. 6.12. In fact, while the TCP connection from an external client (137.204.57.222) to the web server is successful and the HTTP transaction is captured on both interfaces, connection attempts toward the SSH server running on the same host and listening on TCP port 22 are filtered out and do not appear in the DMZ capture.

6.7   Practice: network address translation

Like the case of a packet filter, a Linux box can be configured to act as a NAT middlebox using `iptables` and, specifically, by means of the `nat` table. The flexibility of `iptables` in modifying IP and transport headers is such that it allows the root user to implement a large variety of NAT behaviors. This section illustrates a few examples. As shown in Fig. 6.13, the `nat` table includes three predefined chains:

- PREROUTING, including the NAT rules to be applied to incoming packets before the routing decision is taken, i.e., before the routing table look-up;
- POSTROUTING, including the NAT rules to be applied to outgoing packets after the routing decision is taken;
- OUTPUT, including the NAT rules to be applied to outgoing packets generated by the applications running on the Linux box itself.

The first example shows how to configure a Linux box to act as an outbound basic NAT with static translations in the network set-up of Fig. 6.8. The Linux box interfaces have been configured assigning the private address 192.168.8.254 to the internal interface eth0, while two public addresses 137.204.191.172 and 137.204.191.173 have been assigned to the external interface eth1. The packet forwarding has been enabled in the kernel.

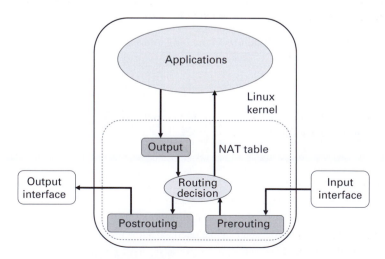

Fig. 6.13   The `nat` table in the Linux kernel

Any outbound NAT configuration requires that, whenever a packet coming from the internal network has to be forwarded to the external network, its private source address is replaced with a public address assigned to the external interface of the middlebox. The kernel knows that a packet is in this situation only after the routing table look-up. Therefore, the outbound NAT rules must be inserted in the POSTROUTING chain specifying that a source NAT (SNAT) operation is to be performed, i.e., the source address must be modified.

Assuming the static translations in Table 6.1, i.e., packets from hosts A and B are always translated into 137.204.191.172 and 137.204.191.173 respectively, the iptables configuration commands to be executed on the Linux box are:

```
iptables -t nat -A POSTROUTING -o eth1 -s 192.168.8.122
        -j SNAT --to-source 137.204.191.172
iptables -t nat -A POSTROUTING -o eth1 -s 192.168.8.4
        -j SNAT --to-source 137.204.191.173
```

At this point, all the traffic from A and B is successfully routed to the external network with the desired address translation, so the hosts that have been contacted on the public network are able to reply to a routable address, as demonstrated by the screen captures shown in Fig. 6.14.

The captures show that the inverse address translation from public to private on reply packets received from the external network is automatically performed by iptables and does not require any additional command from the user. This is because of the intrinsic stateful nature of NAT operations.

The previous configuration limits the NAT capabilities to hosts A and B only: in fact, since packets from a third host on the private network do not match either rule in the POSTROUTING chain, the default policy is applied and the source address is not translated. To increase the number of simultaneous flows allowed through the NAT box, an outbound NAPT behavior must be configured. Therefore, after a flush (-F) command that deletes all the rules previously defined in the POSTROUTING chain, a general NAPT rule can be inserted to force the kernel to translate any address from the private network into one of the public addresses assigned to the Linux box:

```
iptables -t nat -F POSTROUTING
iptables -t nat -A POSTROUTING -o eth1 -s 192.168.8.0/24
        -j SNAT --to-source 137.204.191.172
```

It is possible to test the NAPT features of iptables by generating a TCP connection on each private host toward the same destination and using the same source port. The hping command allows this to be achieved:

```
hping -S -s 1500 -p 80 -c 1 137.204.57.85
```

The command-line options specified above force hping to open a TCP connection (-S) using source port 1500 (-s) toward destination port 80 (-p) on the host with address 137.204.57.85. A single connection is attempted (-c 1). Running exactly the same command as above on both hosts A and B causes the NAT box to modify the source address of outgoing packets belonging to both connections as well as the source port value in TCP headers of the latest connection, so the kernel is able to discriminate the two

Capture on the internal interface (eth0)

No. ▲	Time	Source	Destination	Protocol	Info
1	0.000000	192.168.8.4	155.185.48.5	ICMP	Echo (ping) request
2	0.007263	155.185.48.5	192.168.8.4	ICMP	Echo (ping) reply
3	1.005275	192.168.8.4	155.185.48.5	ICMP	Echo (ping) request
4	1.012023	155.185.48.5	192.168.8.4	ICMP	Echo (ping) reply
5	1.453699	192.168.8.122	155.185.48.5	ICMP	Echo (ping) request
6	1.462012	155.185.48.5	192.168.8.122	ICMP	Echo (ping) reply
7	2.009386	192.168.8.4	155.185.48.5	ICMP	Echo (ping) request
8	2.018311	155.185.48.5	192.168.8.4	ICMP	Echo (ping) reply
9	2.456475	192.168.8.122	155.185.48.5	ICMP	Echo (ping) request
10	2.464426	155.185.48.5	192.168.8.122	ICMP	Echo (ping) reply
11	3.009518	192.168.8.4	155.185.48.5	ICMP	Echo (ping) request
12	3.018844	155.185.48.5	192.168.8.4	ICMP	Echo (ping) reply
13	3.456500	192.168.8.122	155.185.48.5	ICMP	Echo (ping) request
14	3.463957	155.185.48.5	192.168.8.122	ICMP	Echo (ping) reply
15	4.009630	192.168.8.4	155.185.48.5	ICMP	Echo (ping) request
16	4.016505	155.185.48.5	192.168.8.4	ICMP	Echo (ping) reply

Capture on the external interface (eth1)

No. ▲	Time	Source	Destination	Protocol	Info
1	0.000000	137.204.191.173	155.185.48.5	ICMP	Echo (ping) request
2	0.006926	155.185.48.5	137.204.191.173	ICMP	Echo (ping) reply
3	1.005272	137.204.191.173	155.185.48.5	ICMP	Echo (ping) request
4	1.011709	155.185.48.5	137.204.191.173	ICMP	Echo (ping) reply
5	1.453678	137.204.191.172	155.185.48.5	ICMP	Echo (ping) request
6	1.461694	155.185.48.5	137.204.191.172	ICMP	Echo (ping) reply
7	2.009375	137.204.191.173	155.185.48.5	ICMP	Echo (ping) request
8	2.017994	155.185.48.5	137.204.191.173	ICMP	Echo (ping) reply
9	2.456468	137.204.191.172	155.185.48.5	ICMP	Echo (ping) request
10	2.464107	155.185.48.5	137.204.191.172	ICMP	Echo (ping) reply
11	3.009510	137.204.191.173	155.185.48.5	ICMP	Echo (ping) request
12	3.018529	155.185.48.5	137.204.191.173	ICMP	Echo (ping) reply
13	3.456480	137.204.191.172	155.185.48.5	ICMP	Echo (ping) request
14	3.463643	155.185.48.5	137.204.191.172	ICMP	Echo (ping) reply
15	4.009617	137.204.191.173	155.185.48.5	ICMP	Echo (ping) request
16	4.016191	155.185.48.5	137.204.191.173	ICMP	Echo (ping) reply

Fig. 6.14   Simultaneous screen captures on internal and external interface of an outbound basic NAT box with static translations

connections on the external interface. This is testified by the screen captures shown in Fig. 6.15, where for the second connection attempt the original source TCP port (1500) is replaced with a different value (1024).

The same information can be obtained from the kernel NAPT translation table, as shown by the following excerpt:

```
[root@natbox ~]# cat /proc/net/ip_conntrack
tcp [...] src=192.168.8.122 dst=137.204.57.85 sport=1500 dport=80
         src=137.204.57.85 dst=137.204.191.172 sport=80 dport=1500
tcp [...] src=192.168.8.4 dst=137.204.57.85 sport=1500 dport=80
         src=137.204.57.85 dst=137.204.191.172 sport=80 dport=1024
```

The port translation is not performed when it is not necessary, i.e., when different source port values are used by different connections, as happens with parallel HTTP transactions between the same client–server pair. In this case, the NAPT translation table looks like the following excerpt:

```
[root@natbox ~]# cat /proc/net/ip_conntrack
tcp [...] src=192.168.8.122 dst=137.204.57.85 sport=52112 dport=80
         src=137.204.57.85 dst=137.204.191.172 sport=80 dport=52112
tcp [...] src=192.168.8.122 dst=137.204.57.85 sport=52111 dport=80
         src=137.204.57.85 dst=137.204.191.172 sport=80 dport=52111
```

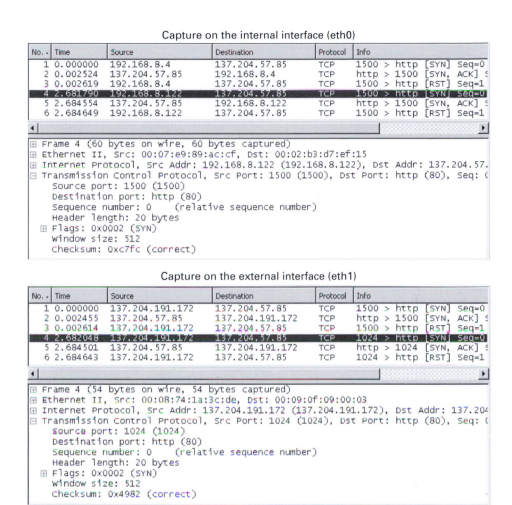

Capture on the internal interface (eth0)

No. ·	Time	Source	Destination	Protocol	Info
1	0.000000	192.168.8.4	137.204.57.85	TCP	1500 > http [SYN] Seq=0
2	0.002524	137.204.57.85	192.168.8.4	TCP	http > 1500 [SYN, ACK] S
3	0.002619	192.168.8.4	137.204.57.85	TCP	1500 > http [RST] Seq=1
4	2.681790	192.168.8.122	137.204.57.85	TCP	1500 > http [SYN] Seq=0
5	2.684554	137.204.57.85	192.168.8.122	TCP	http > 1500 [SYN, ACK] S
6	2.684649	192.168.8.122	137.204.57.85	TCP	1500 > http [RST] Seq=1

```
⊞ Frame 4 (60 bytes on wire, 60 bytes captured)
⊞ Ethernet II, Src: 00:07:e9:89:ac:cf, Dst: 00:02:b3:d7:ef:15
⊞ Internet Protocol, Src Addr: 192.168.8.122 (192.168.8.122), Dst Addr: 137.204.57.
⊟ Transmission Control Protocol, Src Port: 1500 (1500), Dst Port: http (80), Seq: (
      Source port: 1500 (1500)
      Destination port: http (80)
      Sequence number: 0     (relative sequence number)
      Header length: 20 bytes
   ⊞ Flags: 0x0002 (SYN)
      Window size: 512
      Checksum: 0xc7fc (correct)
```

Capture on the external interface (eth1)

No. ·	Time	Source	Destination	Protocol	Info
1	0.000000	137.204.191.172	137.204.57.85	TCP	1500 > http [SYN] Seq=0
2	0.002455	137.204.57.85	137.204.191.172	TCP	http > 1500 [SYN, ACK] S
3	0.002614	137.204.191.172	137.204.57.85	TCP	1500 > http [RST] Seq=1
4	2.682048	137.204.191.172	137.204.57.85	TCP	1024 > http [SYN] Seq=0
5	2.684501	137.204.57.85	137.204.191.172	TCP	http > 1024 [SYN, ACK] S
6	2.684643	137.204.191.172	137.204.57.85	TCP	1024 > http [RST] Seq=1

```
⊞ Frame 4 (54 bytes on wire, 54 bytes captured)
⊞ Ethernet II, Src: 00:08:74:1a:3c:de, Dst: 00:09:0f:09:00:03
⊞ Internet Protocol, Src Addr: 137.204.191.172 (137.204.191.172), Dst Addr: 137.204
⊟ Transmission Control Protocol, Src Port: 1024 (1024), Dst Port: http (80), Seq: (
      Source port: 1024 (1024)
      Destination port: http (80)
      Sequence number: 0     (relative sequence number)
      Header length: 20 bytes
   ⊞ Flags: 0x0002 (SYN)
      Window size: 512
      Checksum: 0x4982 (correct)
```

Fig. 6.15   Simultaneous screen captures on internal and external interface of an outbound NAPT box

The last example of NAT configuration presented here concerns a case of port for-warding, which is basically the dual operation with respect to the outbound NAT. In fact, when a packet from the public network has to be redirected to an internal host, the destination IP address needs to be translated (DNAT) to the corresponding private address. Since this operation must obviously be performed before the routing table look-up takes place, the translation rules are to be inserted in the PREROUTING chain of iptables.

In the example considered here, both private hosts A and B are running an HTTP server that must be reachable from clients on the public network, even though a single public IP address is assigned to the NAT box, i.e., 137.204.191.172. To achieve this, the connection attempts coming from the outside and directed to the internal servers must be differentiated in some way: a possible choice is to configure port forwarding so that connections to port 80 are forwarded to host A, while connections to port 8080 are redirected to port 80 of host B. The corresponding `iptables` rules are:

```
iptables -t nat -A PREROUTING -i eth1 -d 137.204.191.172
        -p tcp --dport 80 -j DNAT --to-destination 192.168.8.122
iptables -t nat -A PREROUTING -i eth1 -d 137.204.191.172
        -p tcp --dport 8080 -j DNAT --to-destination 192.168.8.4:80
```

The correctness of the port forwarding configuration is demonstrated by the screen captures in Fig. 6.16 and by the resulting translation table:

```
[root@natbox ~]# cat /proc/net/ip_conntrack
tcp [...] src=67.171.78.2 dst=137.204.191.172 sport=4624 dport=80
         src=192.168.8.122 dst=67.171.78.2 sport=80 dport=4624
tcp [...] src=67.171.78.2 dst=137.204.191.172 sport=4620 dport=8080
         src=192.168.8.4 dst=67.171.78.2 sport=80 dport=4620
```

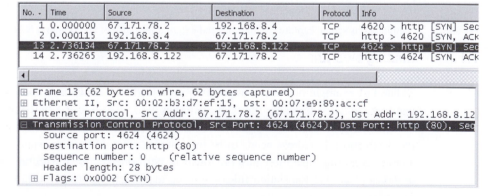

Capture on the external interface (eth1)

No. ▾	Time	Source	Destination	Protocol	Info
1	0.000000	67.171.78.2	137.204.191.172	TCP	4620 > 8080 [SYN] Seq=
2	0.000198	137.204.191.172	67.171.78.2	TCP	8080 > 4620 [SYN, ACK]
13	2.736139	67.171.78.2	137.204.191.172	TCP	4624 > http [SYN] Seq=
14	2.736347	137.204.191.172	67.171.78.2	TCP	http > 4624 [SYN, ACK]

```
⊞ Frame 13 (62 bytes on wire, 62 bytes captured)
⊞ Ethernet II, Src: 00:09:0f:09:00:03, Dst: 00:08:74:1a:3c:de
⊞ Internet Protocol, Src Addr: 67.171.78.2 (67.171.78.2), Dst Addr: 137.204.191.
⊟ Transmission Control Protocol, Src Port: 4624 (4624), Dst Port: http (80), Seq
      Source port: 4624 (4624)
      Destination port: http (80)
      Sequence number: 0      (relative sequence number)
      Header length: 28 bytes
  ⊞ Flags: 0x0002 (SYN)
```

Capture on the internal interface (eth0)

No. ▾	Time	Source	Destination	Protocol	Info
1	0.000000	67.171.78.2	192.168.8.4	TCP	4620 > http [SYN] Seq
2	0.000115	192.168.8.4	67.171.78.2	TCP	http > 4620 [SYN, ACK
13	2.736134	67.171.78.2	192.168.8.122	TCP	4624 > http [SYN] Seq
14	2.736265	192.168.8.122	67.171.78.2	TCP	http > 4624 [SYN, ACK

```
⊞ Frame 13 (62 bytes on wire, 62 bytes captured)
⊞ Ethernet II, Src: 00:02:b3:d7:ef:15, Dst: 00:07:e9:89:ac:cf
⊞ Internet Protocol, Src Addr: 67.171.78.2 (67.171.78.2), Dst Addr: 192.168.8.12
⊟ Transmission Control Protocol, Src Port: 4624 (4624), Dst Port: http (80), Seq
      Source port: 4624 (4624)
      Destination port: http (80)
      Sequence number: 0      (relative sequence number)
      Header length: 28 bytes
  ⊞ Flags: 0x0002 (SYN)
```

Fig. 6.16   Simultaneous screen captures on internal and external interface of a NAT box configured with port forwarding

7 Routing fundamentals and protocols

In Chapter 6, the main functions and architectures of routers have been presented and discussed. Routers are the basic active devices employed in networks and their fundamental task is to forward packets from node to node toward their final destination, taking the most adequate decisions concerning the network path to follow [65]. More specifically, routers perform three main functions: routing table creation and update, table look-up for forwarding decision and physical datagram forwarding from input ports to output ports.

The first of these functions is called *routing* and consists of running algorithms and implementing protocols suitable to the specific routing problem and network topology size [81]. On the one hand, routing algorithms are executed to create and maintain routing tables, once sufficient information about the network topology has been gathered. On the other hand, routing protocols are employed between routers to exchange the network topology information necessary for executing the algorithms.

This chapter illustrates the basic principles of routing, describing the most important algorithms and protocols. The hierarchical approach to the routing problem adopted by the Internet is also discussed, including a brief overview of the most popular IP routing protocols, namely RIP and OSPF. A couple of practical examples of IP routing configurations conclude the chapter.

7.1 Routing algorithms

Routing algorithms are characterized by the way the routing table is created and possibly updated. Generally speaking, such an algorithm should be correct, fair, stable, optimal, fast, simple, and adaptive to network topology changes and to traffic conditions. It is evidently difficult to meet all these requirements with just a single, ideal algorithm, so any real routing algorithm usually provides a trade-off on some of them.

Routing algorithms can be classified as either static or dynamic and as either centralized or distributed.

Static routing means that the routing tables are time-invariant, independent of traffic conditions and modified (typically by hand) only on structural or topological network changes, such as node insertions or removals and link failures. This kind of routing can be either *deterministic*, when the path choice is fixed, or *probabilistic*, when the choice is made as a function of a certain probability distribution over several different alternatives.

Dynamic routing means that routing tables are periodically and automatically updated, for instance to take into account network traffic conditions or topology changes properly.

A *centralized routing* algorithm is performed by a central host that collects state information from all nodes, computes routing tables on behalf of the others and then sends them to all network nodes. The advantage of such a centralized approach is that it guarantees consistent information distribution, so that all routers always have the correct routing information. However, like any centralized scheme, it is affected by issues such as single point of failure and limited scalability for large networks.

Distributed routing algorithms are performed by all nodes, determining their own tables as a function of information that can be either *local*, i.e., related to the node itself without any data exchange with other nodes, or *distributed*, i.e., when it is obtained from other nodes through network links, so that a proper mechanism for these data exchanges has to be set up. This communication mechanism should act in such a way that, on the one hand, it ensures the most consistent information distribution, while on the other hand it keeps the amount of generated overhead as limited as possible.

Among the most commonly used routing algorithms are flooding, deflection routing and shortest path routing.

Flooding is an extremely simple algorithm. Each node retransmits every received packet to all output interfaces. Sooner or later, depending on the length and bit rate of the links, a packet will reach all connected nodes, including the destination. Actually, some modifications are necessary to avoid endless packet propagation in the network. Two rules are then introduced: the first says that a node does not resend a packet to the same link from which it has been received; the second states that a node retransmits a packet only once, discarding retransmitted packets if they come back to the node again. The computation complexity of this algorithm is very low and it can be classified as a dynamic and distributed algorithm.

Flooding is very redundant for point-to-point communications, whereas it is suitable for situations where data must be sent to all nodes (broadcasting), even though some redundancy still persists. An alternative way to route broadcast traffic eliminating redundancies is by using the spanning tree algorithm, as described in Chapter 5.

Deflection routing, known also as *hot potato*, is an algorithm suitable for nodes that have no or small output memories, or when queuing times have to be minimized. On receiving a packet, the node sends it to the output interface with the shortest waiting queue assuming that, after a number of hops, the packet will eventually reach the final destination. It is again a dynamic and distributed algorithm. Of course, since this algorithm does not optimize the path to the real destination of packets, it is likely that packets of a given flow are received out of sequence.

In general, any network topology can be represented as a directed graph, where vertices and edges represent network nodes and links respectively. Given a pair of nodes *A* and *B*, a path from *A* to *B* is a sequence of vertices connected by edges, where *A* and *B* are the two extremes. A graph is said to be connected when a path exists between each node pair. A weighted graph is a graph whose edges are assigned a numerical weight, also called cost or distance. An example of a connected network topology and the corresponding weighted directed graph is shown in Fig. 7.1.

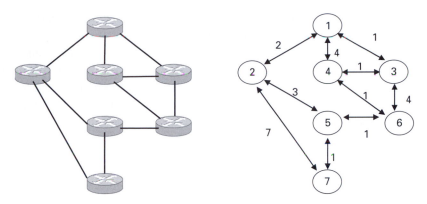

Fig. 7.1 A network topology with seven nodes and ten bidirectional links and the corresponding weighted directed graph

The edge weight can be used to represent the specific cost of network links in all the situations where link characteristics must affect the choice of the path to be followed by packets. Therefore, the cost of a link can be defined in different ways: the physical distance covered by the link, the amount of traffic routed along the link, the inverse of the link bandwidth, to name a few. The cost or distance of a network path is defined as the sum of the costs of all links that are part of the path. If the cost of each edge is assumed to be one, then the path distance is defined in terms of the number of links traversed, which is also called the *hop count*.

Shortest path routing aims at determining the optimal path between each source and destination node pair on the weighted directed graph representing a network. The optimal path is a path with the minimum cost or distance and is called a *shortest path*. Several algorithms to compute the shortest path have been implemented. In general, given a reference destination node, the result obtained after applying such an algorithm on a connected graph is given by two values for each node i of the network:

- D_i, the minimum distance from node i to the reference node;
- n_i, the downstream neighbor, or *next hop*, along the shortest path from node i to the reference node.

These two values are then used to build and update the entries of node i's routing table related to the reference node and the networks connected to it, specifying the distance obtained and the next hop to forward packets to.

The most popular algorithms used by routers to compute the shortest path are the Bellman–Ford and Dijkstra algorithms [82] and these will be explained in the following subsections.

7.1.1 The Bellman–Ford algorithm

To explain the *Bellman–Ford algorithm* better, the centralized version will be illustrated first, followed by the distributed implementation.

Let d_{ij} be the weight of the link connecting node i to node j, where $i \neq j$. If nodes i and j are not directly connected, then assume that $d_{ij} = \infty$. Suppose, without loss of generality, that node 1 is the destination reference node. Let D_i^h be the distance of a shortest path in, at most, h hops between node i and node 1, i.e., the minimum distance for all paths where, at most, h edges are traversed. By definition $D_1^h = 0$ for each h. The goal is to determine the shortest paths from all nodes to the reference node 1 operating step by step by increasing the number h of permitted hops.

Starting with $D_i^0 = \infty$, for each $i \neq 1$, D_i^h can be computed on the hth step as

$$D_i^h = \min_j[d_{ij} + D_j^{h-1}]$$

for each $h \geq 1$ and $i \neq 1$.

The algorithm stops after a finite number of iterations $h_0 \leq N$, where N is the number of nodes. The value $D_i^{h_0}$ obtained at the final step is the shortest distance over all possible paths from node i to node 1, i.e., $D_i^{h_0} = D_i$. Therefore, the shortest path from a generic node $i \neq 1$ to node 1 is given by

$$D_i = \min_j[d_{ij} + D_j]. \tag{7.1}$$

$$n_i = j, \quad \text{such that Eq. (7.1) is satisfied.}$$

If, in the directed graph, there are no loops of negative or zero length, this algorithm always returns only one solution. Figure 7.2 shows an example of application of the Bellman–Ford algorithm.

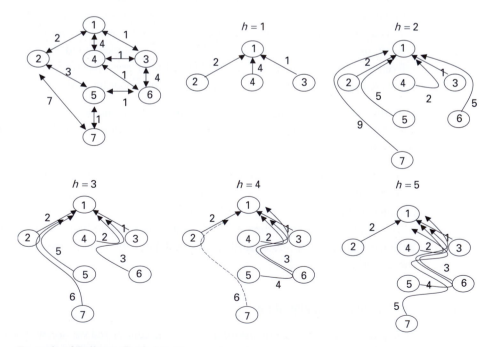

Fig. 7.2   Example of Bellman–Ford algorithm

A routing algorithm based on the Bellman–Ford shortest-path computation can be summarized as follows:

(i) Computation of the shortest paths from each node to all other nodes of the network by a central host, which has to know network topology and link costs to be assigned to the edges of the associated oriented graph.

(ii) Distribution of the results to all nodes for the creation of the respective routing tables; each node makes an association between every possible destination and an output interface on the shortest path toward that destination.

(iii) Every incoming packet is then forwarded by looking at the routing table.

This version of the routing algorithm is centralized, since there is a central host that makes all computations and distributes the results, and static, because shortest paths, once determined, are time invariant.

A distributed version of this algorithm is also available and should be applied whenever the drawbacks of the centralized approach become too significant. Let $N(i)$ be the set of neighbors of node i, i.e., the nodes reachable by node i with a single hop. The initial conditions are $D_i^0 = \infty \ \forall i \neq 1$ and $D_1^0 = 0$, assuming again node 1 as the reference node.

(i) At step h each node i computes

$$D_i^h = \min_{j \in N(i)} [d_{ij} + D_j^{h-1}];$$

(ii) The values of D_i^h are sent from each node i to every neighbor in $N(i)$;

(iii) Each node sets $h = h + 1$ and then iterates from (i).

This algorithm should be run by all nodes in a synchronous way, so that a sort of synchronization mechanism among nodes has to be defined. Of course, whenever changes occur in the network regarding, for instance, topology or link lengths, it is necessary to re-run the algorithm and recompute the shortest paths. However, it is not necessary to set the same time instant for running the algorithm by all nodes and also the initial conditions are not equal for all nodes. This means that the distributed version of the algorithm can also be asynchronous and executed as follows: each node i knows

(i) The last received minimum distance $D_j(t)$ from all nodes $j \in N(i)$ to reference node 1;

(ii) The last evaluation of the minimum distance $D_i(t)$;

(iii) All distances $d_{ij} \ \forall j \in N(i)$, which may change at time instants $t_0, t_1, t_2, \ldots$

Node i periodically updates its shortest path to node 1:

$$D_i(t) = \min_{j \in N(i)} [d_{ij} + D_j(t)].$$

Each node then periodically sends out the updated $D_i(t)$ value to its neighbors. When node i receives from node $j \in N(i)$ a new value for $D_j(t)$, it updates its stored $D_j(t)$ value.

If the following assumptions are true:

(i) Nodes never stop updating minimum distances and transmitting and receiving updated distance values;
(ii) Messages with distance estimations are considered obsolete after a given time period, i.e., a node discards an update message when it is too old;
(iii) There are no loops of negative or zero length in the network;

then a time instant t^* exists such that $D_i(t) = D_i \ \forall t \geq t^*$, where D_i is the minimum distance from node i to node 1.

This means that, in case of link cost modifications, the $D_i(t)$ values computed by this distributed and asynchronous version of the algorithm converge to the correct values.

7.1.2   The Dijkstra algorithm

Another solution to the shortest path computation problem is given by the *Dijkstra algorithm*. Assuming again that node 1 is the reference destination node, during the algorithm execution the set **V** of the N nodes is partitioned into two subsets: node i is in subset **P** if its minimum distance D_i to the reference node has already been computed, whereas subset $\mathbf{P^*} = \mathbf{V} - \mathbf{P}$ includes all nodes whose shortest path to node 1 is still to be found.

Initial conditions are: $\mathbf{P} = \{1\}$, $D_1 = 0$, $D_j = d_{j1} \ \forall j \in \mathbf{P^*}$. The algorithm works iteratively as follows:

(i) Determine the node $i \in \mathbf{P^*}$ such that $D_i = \min_{j \in \mathbf{P^*}} D_j$ and then set $\mathbf{P^*} = \mathbf{P^*} - \{i\}$ and $\mathbf{P} = \mathbf{P} \cup \{i\}$;

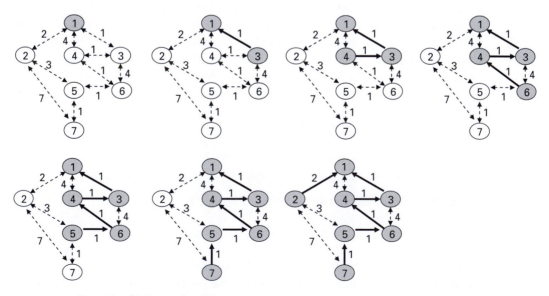

Fig. 7.3   Example of Dijkstra algorithm

(ii) $\forall j \in \mathbf{P}^*$ recompute D_j as $\min[D_j,\ d_{ji} + D_i]$;

(iii) Iterate from (i).

In other words, at each step the algorithm finds the node in $\mathbf{P}^*$ with the minimum distance from any node in $\mathbf{P}$, removes it from $\mathbf{P}^*$ and adds it to $\mathbf{P}$; then the algorithm recomputes the distance of the remaining node in $\mathbf{P}^*$, checking whether the last node added to $\mathbf{P}$ improves the distance of its neighbors, if it is chosen as the next hop. The algorithm stops when all the nodes are in the set $\mathbf{P}$: the distances D_i and the next hops n_i obtained are the correct ones for the shortest path from i to 1. The execution of the Dijkstra algorithm on the same example topology used for the Bellman–Ford case is illustrated in Fig. 7.3.

7.2   Routing protocols

Most routing algorithms require information exchange between nodes: centralized algorithms implemented in a central host have to send routing tables to all network nodes, whereas the distributed algorithms ask for a constant information exchange between nodes so as to keep routing tables updated. Therefore, it is necessary to define *routing protocols* to support these communications, in order to enforce a correct routing function.

The focus in this section is put on the major routing protocols developed for and in service in the Internet. To be suitable to the intrinsic nature of IP networks these protocols are dynamic and distributed and can be grouped into two classes: distance vector and link state.

7.2.1   Distance vector protocols

Distance vector protocols are typically implemented using the distributed and asynchronous version of the Bellman–Ford algorithm. To achieve this, the basic idea is that each node prepares a vector with the known distances to all the other nodes in the network as obtained from the routing table and sends this vector to its neighbors. These protocols were the first ones to be used in the Internet and are usually simple to implement and need very low processing resources.

At the beginning, routing tables have information on the node itself only, at a distance equal to 0. Thus, the first distance vector exchanged includes just these data. Then, following the Bellman–Ford algorithm, such data exchanges allow for the creation of complete routing tables. Since the algorithm is completed in a number of steps at most equal to the number of network nodes, when the network is quite large the convergence time may become significant, leaving the network-wide routing in a temporarily inconsistent state.

Since nodes only exchange information with their neighbors, these protocols may be affected by the *count-to-infinity* problem. Let three nodes, A, B and C, be connected in series, as in Fig. 7.4, and suppose the adoption of a hop count distance, such that $D_{AC} = 2$ and $D_{BC} = D_{BA} = 1$. If the link BC for any reason goes out of service, after a timeout

Fig. 7.4   A typical situation affected by the count-to-infinity problem

period node B deletes the routing information to node C. On the other hand, B learns from A's distance vector that $D_{AC} = 2$ and then it recomputes $D_{BC} = D_{BA} + D_{AC} = 3$. Then B informs A that its new distance to C is $D_{BC} = 3$, forcing A to recompute $D_{AC} = 4$. This process could continue forever, although it is stopped by setting a maximum distance D_{max} such that when $D_{ij} > D_{max}$, node j is assumed to be unreachable.

Split horizon is a very simple technique useful for tackling the count-to-infinity problem. If node A forwards packets to a destination node C through node B, it does not make any sense for node B to try to reach node C through node A. Therefore, it is useless that node A informs B about its distance to C. This modification to the routing protocol implies that a node now has to send different distance vectors to different neighbor nodes, by properly selecting information from its table. This makes the distance vector creation process a little more complicated.

Triggered update is a further improvement aimed at reducing the time required to converge to the shortest path solution. It deals with the timing for sending updated distance vectors to the neighbors. Besides the periodic updates sent by each node to its neighbors, the triggered update technique requires that, if there is any change in the routing table, a node immediately sends an updated distance vector. This behavior reduces the chance of spreading wrong routing information caused by outdated distance vectors.

Although solutions such as split horizon and triggered update improve the performance of a distance vector protocol, convergence problems still persist, especially when cycles are present in the network topology. A possible way to drastically overcome this issue is to include the complete list of nodes traversed to reach each destination in the distance vector sent to neighbors, which is now called *path vector*. A router is, thus, able to select only the valid elements in the path vectors received by the neighbors by ignoring the destinations with a path where the router itself is already present. This solution eliminates the risk of cycles in routing paths, although it requires additional overhead due to the increased amount of information to be exchanged.

7.2.2   Link state protocols

An alternative to distance vector protocols is represented by *link state* protocols. The basic characteristic of this kind of routing protocols is that each node tries to build an image of the complete network topology. Then, using this topology, each node computes its own routing table, typically applying the Dijkstra algorithm. Therefore, such routing protocols must implement a communication mechanism allowing network nodes to learn the entire network topology in a distributed fashion.

To this purpose, each node or router builds a specific packet, called a *link state packet* (LSP), which contains the list of its neighbors and the cost of the link to reach them. This packet is then sent to all the routers in the network. After receiving the LSPs from all the other nodes, a router is eventually able to reconstruct the entire network topology, including link costs, and apply the shortest path algorithm for any possible destination.

Since the correct distribution of LSPs is critical for building a consistent network topology, it is very important to avoid the situation where different routers have different collections of LSPs because this would imply different route computations that could lead to wrong routing decisions. In addition, the distribution of LSPs should not waste too much bandwidth, so this process must be carefully organized.

The distribution of LSPs to other nodes cannot be simply based on current routing tables, because these are supposed to be updated during this process. The distribution can then be performed using a flooding mechanism. Each node keeps copies of the most recent LSPs received, so that it retransmits only the LSPs not already received. An LSP could be delayed in the network so that it can finally reach a router after a more recent LSP. To avoid this problem, which may cause outdated LSPs to overwrite correct information with a consequent routing instability, it is necessary to mark the LSPs sent out by each node sequentially. Therefore, two fields are introduced in the LSP header: a sequence number and a time-to-live value, which is decreased each time the LSP crosses a node.

As a consequence, a valid LSP should include the following fields: source node address, sequence number, age, list of neighbors and their distance.

7.2.3   Distance vector, path vector or link state?

Distance vector, path vector and link state routing protocols are all available in IP networks. But when is it advisable to use one kind instead of another? To answer this question, let us consider the pros and cons of either approach.

  (i) Distance vector protocols:
      (a) Are simple;
      (b) Require limited computational and storage capabilities;
      (c) Suffer from slow convergence, especially in the case of large topologies with cycles;
 (ii) Path vector protocols:
      (a) Are relatively more complex;
      (b) Eliminate the problem of cycles in the topology;
      (c) Allow selection of a path based on the list of traversed nodes;
      (d) May still suffer from slow convergence problems;
(iii) Link state protocols:
      (a) Are more complex and require significant computational capabilities, since the Dijkstra algorithm must be executed each time the topology is updated;
      (b) Require significant storage space, as the LSPs from every network node must be kept;

(c) May generate significant overhead due to the LSP flooding;

(d) Are fast to converge and do not have problems with cycles, even in large topologies;

(e) Offer advanced network-wide management capabilities, since each router knows the entire topology.

All the three options have advantages and drawbacks and the optimal choice must be made based on a trade-off that takes into account the specific situation to deal with. For instance, distance vector protocols should be adopted in small and simple topologies made by low-cost routers, whereas larger networks with complex topologies and powerful routers should opt for the link state family. Finally, when topology changes are not very frequent and the optimal routing depends mainly on which nodes are to be traversed rather than a distance parameter, the choice of a path vector solution could be advisable.

7.3   Routing in the Internet

Since IP is a protocol for packet-switched, connectionless networks, the execution of a dynamic and distributed routing algorithm and the implementation of a suitable routing protocol are very critical functions that must be correctly performed at the whole Internet level. However, when a packet-switched network is extremely large, as in the case of the global Internet, managing routing tables that include all the possible destinations in the whole network is a very complex task. To understand the real dimension of this problem, consider that in the first half of 2009 routers in the Internet core had to deal with almost 300 000 entries in their routing tables[1] and this number keeps growing.

Therefore, the only feasible way to implement IP routing is to use a hierarchical approach. The Internet is a collection of routers and networks interconnected by means of the IP protocol. The entire network is then partitioned into several subsets that we can generically call "areas." Routers within each area are responsible for routing packets between networks internal to the same area: this drastically reduces the size of the tables of internal routers. As regards all the other possible destinations, internal routers are instructed to forward packets addressed outside the area to some special routers, placed at the borders, which know the external topology and have larger routing tables including all the possible network prefixes. Border routers are then responsible for routing packets between different areas. In principle, area partitions can be applied iteratively, originating multiple levels in the routing hierarchy and further reducing the number of border routers that need to maintain a complete routing table.

Another advantage of hierarchical routing is that the routing process inside an area is independent of the routing processes within other areas and can also be decoupled from the inter-area routing. This means that different areas are allowed to adopt different internal routing protocols, whereas a suitable common external routing protocol must be adopted between areas.

---

[1] Source: *BGP Routing Table Analysis Reports*, http://bgp.potaroo.net/.

According to the terminology used by the Internet routing, the global network is partitioned at the highest hierarchical level into subsets called *autonomous systems* (ASs), which generally represent domains of networks and routers under a given administrative entity. Routing protocols internal to an autonomous system are called *interior gateway protocols* (IGP), whereas the ones used between different autonomous systems are called *exterior gateway protocols* (EGP). A second-level partitioning could be applied within a given autonomous system, which is then split into *routing areas* (RA). However, this is not mandatory, as this feature depends on the interior protocol used within the autonomous system.

An example of Internet-like hierarchical structure, with a few interconnected autonomous systems, and the resulting topology as seen by the exterior routing protocol is shown in Figs. 7.5 and 7.6. Note how the routing problem is simplified when the topology to be considered is limited to routers and networks within a single autonomous system (or routing area) or to border routers only.

Among the most commonly used IGPs are RIP, OSPF and IS–IS, whereas the EGP currently used in the Internet is BGP. The RIP [83] and OSPF [84] protocols are very popular interior protocols and will be illustrated in the next sections. *Intermediate system to intermediate system* (IS–IS) is a link state protocol used for interior routing and represents an alternative to OSPF. It has been standardized in documents ISO/IEC 10589:1992 and ISO/IEC 10589:2002(E) [85] and IETF described it in RFCs 1142 [86] and 1195 [87]. Being an ISO standard, IS–IS was designed to route a general layer-3 packet format, making it more flexible to work with different network protocols than OSPF. The working principles of IS–IS are similar to those of OSPF. Although IS–IS is not as popular as OSPF, it is successfully employed within some autonomous systems, especially in the case of large Internet service providers.

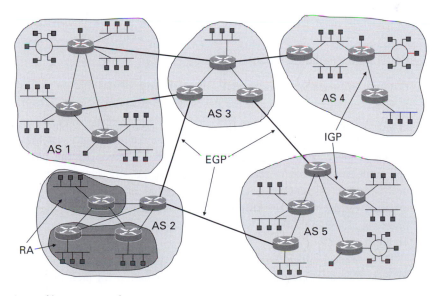

Fig. 7.5   A set of interconnected autonomous systems

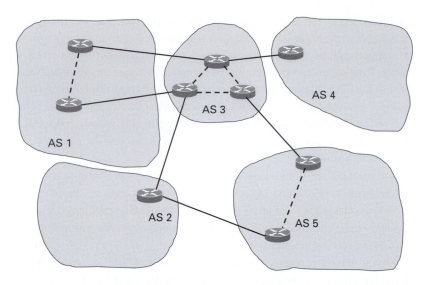

Fig. 7.6   The same network topology shown in Fig. 7.5 as seen by the exterior routing protocol

Border Gateway Protocol (BGP) version 4 is the external protocol currently used for routing between different Internet autonomous systems and is described in RFC 1771 [88]. It is a path vector protocol and, as such, it is appropriate for the kind of decisions made at the highest routing level. Indeed, when the administration of an autonomous system must decide what kind of external traffic is allowed to traverse its networks and which peering relationships must be established with other autonomous systems, the ability to know the exact path followed by each route proves to be very useful. Therefore, BGP routing tables are populated based mainly on routing policies rather than optimal shortest path computations.

7.3.1   Routing information protocol

Routing information protocol (RIP) has been one of the first and most widely used Internet routing protocols since the 1980s. The first version was described in RFC 1058, followed by a second version in RFC 2453 [83]. It is a distance vector protocol and, as such, it can be used as an interior protocol in small autonomous systems.

The distance between two hosts or networks is given by the number of links a datagram has to cross to reach the destination, i.e., the hop count metric is used. A destination is assumed unreachable when it has an infinite distance, which is represented by the value 16. This value has been chosen to solve the count-to-infinity problem, although it obviously limits the size of the RIP network to paths shorter than 16 hops.

This simple protocol basically uses two types of messages: request and response. *Request* messages are used to ask neighbors explicitly for their distance vectors, e.g., just after a router boots up. *Response* messages are usually employed to transmit routing information in the form of distance vectors. A response message is generated and sent out with updated information:

  (i) Periodically;
 (ii) As a reply to a request message;
(iii) Whenever a change in the routing information occurs (triggered update).

Periodic response messages are typically sent every 30 seconds, with a variation of one to five seconds to avoid storms of update packets sent at the same time.

    Besides the usual fields, such as destination address, distance and next hop, each entry of the RIP routing table contains also information about its lifetime, so that destinations that have not been updated within a given period, called *timeout* (typically 180 seconds), are considered stale and their distance is set to infinite. After an additional expiration time, called *garbage-collection time* (typically 120 seconds), such entries are completely removed from the table.

    When a RIP router A receives a response message from a neighbor B, it first checks the validity of data included in the distance vector. Then it extracts only the entries with a distance value smaller than 16. The distance obtained from each of the entries considered is increased by one, to include the additional hop to reach the neighbor B, and is compared with the current value in the routing table corresponding to the same destination address. If such an entry is not present in the routing table, it means that A was not aware of this destination and the entry is created from scratch with the distance value just computed and including B as a next hop. Otherwise, if the destination is already known by A, the routing table is updated only in two cases:

  (i) When the old distance value is higher than the new one, meaning that a shorter path has been found;
 (ii) When the old distance value is different from the new one but the next hop is the same, meaning that the neighbor has updated its distance.

Whenever a table entry is created, updated or simply confirmed, the timeout is restarted.

    The RIP applies the split horizon technique, so that response messages sent to different interfaces are, in general, different. It also applies triggered update, in which case it is not necessary to report all distance vector entries in the response message but only the updated ones. The RIP uses UDP as the transport protocol with both source and destination port number 520.

    The RIP version 1 is an unsecure protocol, since it accepts updates from anyone without any authentication procedure or access control: this means that it is prone to attacks aimed at forcing the router to forward packets along a given path, e.g., to intercept these packets or to simply break network connectivity. In addition, RIPv1 is not capable of associating a netmask to a destination IP address, meaning that only the classful addressing scheme is supported. The RIP version 2 improves these aspects, as it is enhanced to provide specific fields in the request and response messages to include authentication data and netmask values.

7.3.2   Open shortest path first

Open shortest path first (OSPF) is a very popular and flexible link state protocol, widely used in the Internet as an interior routing protocol. It became a standard in its second version, described in RFC 2328 [84], and was designed with the objective of overcoming most of the problems previously encountered with the use of RIP. In particular, besides the adoption of a link state protocol that is more suitable for large networks with complex topologies, some of the peculiar aspects of OSPF include:

- The capability to add a third level in the Internet routing hierarchy;
- An efficient exchange of routing information with relation to the intrinsic nature of different network types;
- The ability to set the cost of each link to a desired value;
- The capability of load balancing over multiple minimum distance paths;
- An authentication procedure to guarantee information exchange with authorized routers only.

Open shortest path first enforces hierarchical routing at the single autonomous system level, allowing the definition of multiple routing areas. Routers internal to an area (*internal routers*) are required to exchange information within and know the topology of the routing area only. This means that a smart partitioning of the autonomous system into routing areas allows a significant reduction of the processing burden on internal routers for the computation of the Dijkstra algorithm. The requirement to guarantee complete connectivity within the autonomous system is that each area must be connected to a common area called the *backbone*, or *area 0*. The backbone is responsible for routing information exchange between different areas, similarly to what exterior protocols do between different autonomous systems.

Therefore, each non-zero area must have at least a router (*area border router*) connected to a router in the backbone (*backbone router*), either through a direct physical connection or through a virtual link traversing other routers. In addition, at least one router within the autonomous system must be responsible for redistributing routing information exchanged with other autonomous systems using an exterior protocol. This kind of router (*AS-boundary router*) represents the gateway to access the rest of the Internet. An example of OSPF topology and router classification is shown in Fig. 7.7.

Open shortest path first has been designed to deal efficiently with networks showing different physical characteristics, such as point-to-point links (e.g., serial connections), broadcast multiple-access networks (e.g., Ethernet, Wi-Fi) and non-broadcast multiple-access networks (e.g., ATM, frame relay). In fact, different network types impose different methods to establish full adjacency relationships, which are required to exchange routing information. Adjacency is trivial in point-to-point networks, where the neighbors at the two endpoints of the link must exchange routing information.

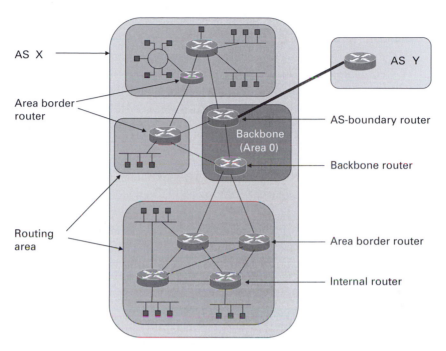

Fig. 7.7   Example of OSPF topology and router classification

It is different when N routers are attached to the same broadcast network: here each router is a neighbor to every other router, for a total of $N(N-1)/2$ connections. However, to reduce the overhead generated by the routing information exchange, each router becomes fully adjacent to one neighbor only, which is also responsible for distributing the link state packet describing the network. This particular router is called the *designated router* (DR) and is chosen among the N routers connected to the network through an election process. For reliability purposes, a *backup designated router* (BDR) is elected as well. Since link state packets are exchanged only between fully adjacent routers, the election of the DR allows the routing protocol overhead to be limited because only $N-1$ adjacencies are established instead of $N(N-1)/2$. Moreover, the broadcast nature of the shared medium allows further reduction of the OSPF overhead, since link state packets can be transmitted using a multicast address, meaning that a single packet sent by the DR reaches all its adjacent routers simultaneously.

When N routers are connected to a point-to-multipoint network through a non-broadcast medium, two different choices are possible, depending on the lower-layer protocols and configurations used. The first option consists of establishing full adjacency between any pair of routers, emulating the presence of $N(N-1)/2$ point-to-point connections. The alternative is to emulate the behavior of a broadcast network by electing a DR and establishing $N-1$ full adjacencies with the other routers.

Another OSPF feature that adds flexibility to the router configuration is represented by the possibility for the router administrator to define the cost of each interface based on the most appropriate parameter. This allows the administrator to give preference to paths with particular characteristics as well as to create multiple minimum-cost paths over which traffic can be balanced. Furthermore, OSPF adopts an elaborate and smart methodology to define the distance value to be assigned to routing information, based on the origin of such information. For instance, routes originated from outside an area are assigned a higher distance value than routes generated internally, with the beneficial effect of keeping local traffic within the area. Also, routes redistributed from other protocols are given higher distance values depending on the origin, with the objective of penalizing less reliable protocols, such as RIP, or manually configured error-prone static routes.

Information exchanged between OSPF routers is carried by OSPF packets, which are encapsulated directly in IP packets, so no transport protocol is used. Different types of OSPF packets are defined by the standard, with a common header including such information as router identifier, area number and authentication data. *Hello* packets are periodically sent on each interface to perform several functions, such as to monitor the link availability, to discover neighbors and maintain connectivity and adjacency with them, to elect DR and BDR. Specific timeouts are used to decide when a neighbor is not reachable any more.

When OSPF packets are used to exchange actual routing information, they carry messages called *link state advertisements* (LSAs). Depending on the nature of the information exchanged, an LSA can be:

- A *router-LSA*, originated from any router and including link state information of its interfaces;
- A *network-LSA*, originated from a DR and including information about the respective multiple-access network, such as the list of connected routers;
- A *summary-LSA*, originated from an area border router and including routing information for reaching destinations on other areas;
- An *AS-boundary-router-summary-LSA*, originated from an area border router and including routing information for reaching AS-boundary routers;
- An *AS-external-LSA*, originated from an AS-boundary router and including routing information for reaching destinations on other autonomous systems.

The collected LSAs are stored by a router in the *link state database*, which is used to build an image of the network topology and apply the shortest path algorithm. Routers in the same area must keep the same link state database to maintain consistent routing information. Therefore, after an adjacency has been established, *database description* packets are exchanged between the two routers so that they are able to check for possible differences. These packets do not carry the complete LSAs, but only summary information that allows one to verify whether the two databases are aligned. This information includes the LSA type, its age, the identifier of the router that originated it and the sequence number. In case of discrepancies between the two databases, the missing updated LSAs are explicitly requested through *link state request* packets.

In response to a link state request or whenever the state of a link changes, e.g., because of topology or cost modifications, the relevant LSAs are distributed to all the routers by means of *link state update* packets sent through a flooding process. Link state update packets are also periodically generated by routers to increase the robustness of the routing information exchange process. However, such periodic updates must not be too frequent to avoid large protocol overhead. When a router forwards a link state update to its adjacent neighbors, it expects a *link state acknowledgment* packet, otherwise the update is retransmitted. This improves the flooding mechanism by adding reliability to the communication process.

What we have illustrated here are the basic principles of OSPF. The protocol is much more elaborate and a lot of details have been neglected. Further information can be obtained from several textbooks available that cover this topic or directly from the relevant RFCs.

7.4 Practice: RIP configuration

The ability of the Linux kernel to act as an IP router, with particular reference to the forwarding and filtering functions, has already been shown in Chapter 6. What is still missing is a demonstration of how a Linux box can become a fully functional IP router, also implementing the most important Internet routing protocols. In particular, this section explains how to set up RIP routing in a simple network made by Linux-based routers. A number of open-source implementations of the IP routing protocols are available for Linux systems, among which XORP [89] and Quagga [90] are the most popular ones. In this chapter, the Quagga routing suite will be used, although in principle XORP would have been an equally valid choice.

Quagga implements several IP routing protocols, such as RIPv1, RIPv2, OSPF, BGP and IPv6 routing protocols. Each protocol is activated through a specific service (or daemon) that, once started, enables both the protocol operations and a command-line configuration interface. Quagga routing daemons implementing RIP, OSPF and BGP are named `ripd`, `ospfd` and `bgpd` respectively. These daemons are specialized in running the respective protocols and collect all the information required to update the routing table, which is managed by the Linux kernel. However, they are not allowed to change the kernel table directly, as a sort of coordinating sublayer is required. In fact, the same box could be configured to implement multiple routing protocols, e.g., a router acting as an OSPF AS-boundary router is supposed also to run BGP and to redistribute routes between the two protocols. For this reason, Quagga requires an additional daemon called `zebra` to be running that acts as a collector of the routing updates from the different protocols and modifies the kernel table accordingly. The `zebra` daemon is also responsible for route redistribution between different protocols. The architecture of the Quagga suite and the related module interactions are shown in Fig. 7.8.

The reference network topology and the related IP address plan chosen for the case study discussed in this section are shown in Fig 7.9. The figure also shows the name of the router interfaces and the IP addresses to be assigned to them. The two hosts have been configured according to the address plan by means of the usual `ifconfig` command, whereas the `route` command has been used to set routers *A* and *D* as default gateways for the leftmost and rightmost host respectively. As explained in Chapter 6, the packet

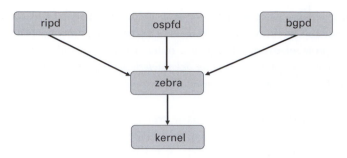

Fig. 7.8   Architecture of the Quagga software routing suite

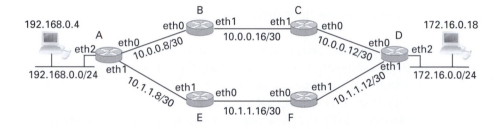

Router	eth0	eth1	eth2
A	10.0.0.9	10.1.1.9	192.168.0.254
B	10.0.0.10	10.0.0.17	N/A
C	10.0.0.13	10.0.0.18	N/A
D	10.0.0.14	10.1.1.14	172.16.0.254
E	10.1.1.17	10.1.1.10	N/A
F	10.1.1.18	10.1.1.13	N/A

Fig. 7.9   The network case study and the related IP address plan

forwarding function must be enabled in the kernel of all the six Linux routers with the `sysctl` command:

```
[root@RouterA ~]# sysctl -w net.ipv4.ip_forward=1
net.ipv4.ip_forward = 1
```

Once the forwarding engine has been prepared, the `zebra` routing daemon must be started. Depending on the specific Linux distribution used, start-up scripts may be available to help the root user to activate the daemon in the most correct way:

```
[root@RouterA ~]# /etc/init.d/zebra start
Starting zebra: Nothing to flush.                    [  OK  ]
```

As any other Quagga routing daemon, once started, `zebra` also enables a command-line configuration interface, which can be accessed via a telnet connection directed to TCP port 2601 of the loopback interface:

```
[root@RouterA ~]# telnet 127.0.0.1 2601
Trying 127.0.0.1...
Connected to RouterA (127.0.0.1).
Escape character is '^]'.

Hello, this is Quagga (version 0.98.6).
Copyright 1996-2005 Kunihiro Ishiguro, et al.

User Access Verification

Password: ********
RouterA-zebra>
```

Any command-line interface used to configure Quagga is protected by two levels of restricted access: the first password is required when accessing the configuration interface via telnet, as shown above; the second one is required when the user types the `enable` command to enter the privileged mode that allows to actually configure the router:

```
RouterA-zebra> enable

Password: ********
RouterA-zebra#
```

From now on, all the commands entered in any Quagga configuration interface are typed in privileged mode, as revealed by the # sign in the command prompt. It is worth noting that the two-level restricted access and, in general, the syntax used by Quagga command-line interfaces recall the same tools available in commercial routers, such as Cisco™ products.

Being responsible for updating the kernel configuration, `zebra` can be used to assign IP addresses to the router interfaces and add static routes to the kernel table. Since the Linux routers must still be configured according to the address plan of Fig 7.9, an example of the sequence of commands to do this on router A using `zebra` is as follows:

```
RouterA-zebra# configure terminal
RouterA-zebra(config)# interface eth0
RouterA-zebra(config-if)# ip address 10.0.0.9/30
RouterA-zebra(config-if)# exit
RouterA-zebra(config)# interface eth1
RouterA-zebra(config-if)# ip address 10.1.1.9/30
RouterA-zebra(config-if)# exit
RouterA-zebra(config)# interface eth2
RouterA-zebra(config-if)# ip address 192.168.0.254/24
RouterA-zebra(config-if)# exit
RouterA-zebra(config)# exit
RouterA-zebra#
```

After entering configuration mode (`configure terminal`), each interface must be selected (e.g., `interface eth0`) and the corresponding IP address must be specified (e.g., `ip address 10.0.0.9/30`). The `exit` command allows one to go back to the previous configuration level.

In the same way, all the other routers must be assigned their IP addresses according to the address plan.

The next step is to start the `ripd` daemon and enter its configuration interface, using TCP port 2602 this time:

```
[root@RouterA ~]# /etc/init.d/ripd start
Starting ripd:                              [  OK  ]
[root@RouterA ~]# telnet 127.0.0.1 2602
Trying 127.0.0.1...
Connected to RouterA (127.0.0.1).
Escape character is '^]'.

Hello, this is Quagga (version 0.98.6).
Copyright 1996-2005 Kunihiro Ishiguro, et al.

User Access Verification

Password: ********
RouterA-ripd> enable

Password: ********
RouterA-ripd#
```

For security reasons, RIPv2 must be configured. Therefore, the first thing to be defined is the authentication key to be used to exchange distance vectors with other routers:

```
RouterA-ripd# configure terminal
RouterA-ripd(config)# key chain RIPkey
RouterA-ripd(config-keychain)# key 1
RouterA-ripd(config-keychain-key)# key-string testRIPkey
RouterA-ripd(config-keychain-key)# exit
RouterA-ripd(config-keychain)# exit
RouterA-ripd(config)#
```

The authentication string is `testRIPkey` and it is defined as the key number one in the key chain named `RIPkey`. This procedure allows the definition of multiple keys in each key chain to enhance authentication robustness. In addition, different key chains can be defined so that they can be used to authenticate different neighbors.

Now the authentication type and the key chain to be used on each interface must be specified. Assuming that the same key chain is used on each interface of router *A* with an MD5 encryption, the command sequence to be entered is:

```
RouterA-ripd(config)# interface eth0
RouterA-ripd(config-if)# ip rip authentication mode md5
RouterA-ripd(config-if)# ip rip authentication key-chain RIPkey
RouterA-ripd(config-if)# exit
```

```
RouterA-ripd(config)# interface eth1
RouterA-ripd(config-if)# ip rip authentication mode md5
RouterA-ripd(config-if)# ip rip authentication key-chain RIPkey
RouterA-ripd(config-if)# exit
RouterA-ripd(config)# interface eth2
RouterA-ripd(config-if)# ip rip authentication mode md5
RouterA-ripd(config-if)# ip rip authentication key-chain RIPkey
RouterA-ripd(config-if)# exit
RouterA-ripd(config)#
```

It is advised to enable the authentication on interface eth2 as well, even though no RIP routers are present on the connected LAN. This is to prevent malicious users connected to the LAN who are sending fake RIP responses from compromising the routing table integrity. Similar key definitions must be configured on the other routers, making sure to use the same key strings on the interfaces connecting two neighbors.

The final step required to complete the router configuration is to activate the RIPv2 routing process and enable it on each IP network that must be advertised by the protocol:

```
RouterA-ripd(config)# router rip
RouterA-ripd(config-router)# version 2
RouterA-ripd(config-router)# network 10.0.0.8/30
RouterA-ripd(config-router)# network 10.1.1.8/30
RouterA-ripd(config-router)# network 192.168.0.0/24
RouterA-ripd(config-router)# exit
RouterA-ripd(config)# exit
RouterA-ripd#
```

After similar configurations have been set up on the other routers, the RIP protocol starts exchanging routing information. When the routing algorithm converges, RIP response packets carrying distance vectors are periodically exchanged between neighbors. An example of a response packet generated by router A as captured on the link AB is shown in Fig. 7.10. The packet contains a distance vector with four destinations and the corresponding distance from A in terms of hop count. It is interesting to compare such a distance vector with A's RIP routing table:

```
RouterA-ripd# show ip rip
Codes: R - RIP, C - Connected, S - Static, O - OSPF, B - BGP
Sub-codes:
       (n) - normal, (s) - static, (d) - default, (r) - redistribute,
       (i) - interface

       Network            Next Hop      Metric From          Tag Time
C(i)   10.0.0.8/30        0.0.0.0           1 self            0
R(n)   10.0.0.12/30       10.0.0.10         3 10.0.0.10       0 02:52
R(n)   10.0.0.16/30       10.0.0.10         2 10.0.0.10       0 02:52
C(i)   10.1.1.8/30        0.0.0.0           1 self            0
R(n)   10.1.1.12/30       10.1.1.10         3 10.1.1.10       0 02:47
R(n)   10.1.1.16/30       10.1.1.10         2 10.1.1.10       0 02:47
R(n)   172.16.0.0/24      10.0.0.10         4 10.0.0.10       0 02:52
C(i)   192.168.0.0/24     0.0.0.0           1 self            0
RouterA-ripd#
```

```
⊞ Frame 139 (166 bytes on wire, 166 bytes captured)
⊞ Ethernet II, Src: 00:0e:0c:69:17:40, Dst: 01:00:5e:00:00:09
⊞ Internet Protocol, Src Addr: 10.0.0.9 (10.0.0.9), Dst Addr: 224.0.0.9 (224.0.0.9)
⊞ User Datagram Protocol, Src Port: router (520), Dst Port: router (520)
⊟ Routing Information Protocol
    Command: Response (2)
    Version: RIPv2 (2)
    Routing Domain: 0
  ⊞ Authentication: Keyed Message Digest
  ⊟ IP Address: 10.1.1.8, Metric: 1
      Address Family: IP (2)
      Route Tag: 0
      IP Address: 10.1.1.8 (10.1.1.8)
      Netmask: 255.255.255.252 (255.255.255.252)
      Next Hop: 0.0.0.0 (0.0.0.0)
      Metric: 1
  ⊟ IP Address: 10.1.1.12, Metric: 3
      Address Family: IP (2)
      Route Tag: 0
      IP Address: 10.1.1.12 (10.1.1.12)
      Netmask: 255.255.255.252 (255.255.255.252)
      Next Hop: 0.0.0.0 (0.0.0.0)
      Metric: 3
  ⊟ IP Address: 10.1.1.16, Metric: 2
      Address Family: IP (2)
      Route Tag: 0
      IP Address: 10.1.1.16 (10.1.1.16)
      Netmask: 255.255.255.252 (255.255.255.252)
      Next Hop: 0.0.0.0 (0.0.0.0)
      Metric: 2
  ⊟ IP Address: 192.168.0.0, Metric: 1
      Address Family: IP (2)
      Route Tag: 0
      IP Address: 192.168.0.0 (192.168.0.0)
      Netmask: 255.255.255.0 (255.255.255.0)
      Next Hop: 0.0.0.0 (0.0.0.0)
      Metric: 1
```

Fig. 7.10   Screen capture of a RIP response packet generated by router *A* on link *AB*

The destinations included in the distance vector sent by *A* on interface `eth0` are only half of the prefixes present in its routing table. Specifically, only the prefixes reachable through *E* (i.e., where the next hop is 10.1.1.10) or directly attached to *A* and not to *B* are sent on the link *AB*. This is clearly a consequence of the split horizon technique adopted. The current value of the timeout associated with each entry is also clearly visible in the RIP routing table.

The `route` command allows one to verify that the routing information collected through RIP has been successfully used by `zebra` to update the kernel routing table:

```
[root@RouterA ~]# route -n
Kernel IP routing table
Destination   Gateway      Genmask         Flags Metric Iface
10.1.1.12     10.1.1.10    255.255.255.252 UG    3      eth1
10.1.1.8      0.0.0.0      255.255.255.252 U     0      eth1
10.0.0.12     10.0.0.10    255.255.255.252 UG    3      eth0
10.0.0.8      0.0.0.0      255.255.255.252 U     0      eth0
10.0.0.16     10.0.0.10    255.255.255.252 UG    2      eth0
10.1.1.16     10.1.1.10    255.255.255.252 UG    2      eth1
172.16.0.0    10.0.0.10    255.255.255.0   UG    4      eth0
192.168.0.0   0.0.0.0      255.255.255.0   U     0      eth2
```

From the test topology shown in Fig. 7.9, it is apparent that two equal-cost paths are present between the two LANs, namely *ABCD* and *AEFD*, but only the former is actually used to route packets, as confirmed by a traceroute executed on the leftmost host toward the other one:

```
[root@host4 ~]# traceroute -n 172.16.0.18
traceroute to 172.16.0.18 (172.16.0.18), 30 hops max, 40 byte
 1  192.168.0.254  7.137 ms    0.216 ms    0.086 ms
 2  10.0.0.10  3.652 ms    3.496 ms    0.152 ms
 3  10.0.0.18  5.839 ms    5.567 ms    0.234 ms
 4  10.0.0.14  5.041 ms    4.706 ms    0.303 ms
 5  172.16.0.18  7.002 ms    6.608 ms    2.202 ms
[root@host4 ~]#
```

The two paths are totally equivalent to RIP and the reason why one is used instead of the other is simply a consequence of the order in which the routers enabled the RIP process.

To test the reconfiguration capability of RIP, a ping session is started between the two hosts. During this session, the link *CD* is forced to go down, causing RIP to trigger distance vector updates and to recompute the shortest path. The amount of time required to converge again and restore complete connectivity can be estimated by analyzing the ping trace, in particular the evolution of the sequence number:

```
root@host4 ~]# ping 172.16.0.18
PING 172.16.0.18 (172.16.0.18) 56(84) bytes of data.
64 bytes from 172.16.0.18: icmp_seq=1 ttl=60 time=0.516 ms
64 bytes from 172.16.0.18: icmp_seq=2 ttl=60 time=0.426 ms
64 bytes from 172.16.0.18: icmp_seq=3 ttl=60 time=0.431 ms
64 bytes from 172.16.0.18: icmp_seq=4 ttl=60 time=0.423 ms
64 bytes from 172.16.0.18: icmp_seq=5 ttl=60 time=0.436 ms
64 bytes from 172.16.0.18: icmp_seq=169 ttl=60 time=7.77 ms
64 bytes from 172.16.0.18: icmp_seq=170 ttl=60 time=0.566 ms
64 bytes from 172.16.0.18: icmp_seq=171 ttl=60 time=0.549 ms
64 bytes from 172.16.0.18: icmp_seq=172 ttl=60 time=0.577 ms
64 bytes from 172.16.0.18: icmp_seq=173 ttl=60 time=0.534 ms
64 bytes from 172.16.0.18: icmp_seq=174 ttl=60 time=0.603 ms
64 bytes from 172.16.0.18: icmp_seq=175 ttl=60 time=0.556 ms

--- 172.16.0.18 ping statistics ---
175 packets transmitted, 12 received, 93% packet loss, time 174019ms
rtt min/avg/max/mdev = 0.423/1.116/7.777/2.009 ms
```

During the routing outage, 163 ICMP packets are lost before connectivity is restored. Since ping packets are sent every second, it is possible to estimate that more than 2.5 minutes are needed by RIP to converge again. The alternative path is eventually chosen, as demonstrated by repeating the traceroute:

```
[root@host4 ~]# traceroute -n 172.16.0.18
traceroute to 172.16.0.18 (172.16.0.18), 30 hops max, 40 byte
 1  192.168.0.254  0.119 ms    0.085 ms    0.094 ms
 2  10.1.1.10 (10.1.1.10)  0.201 ms    0.169 ms    0.173 ms
 3  10.1.1.18 (10.1.1.18)  0.357 ms    0.314 ms    0.344 ms
 4  10.1.1.14 (10.1.1.14)  0.512 ms    0.437 ms    0.386 ms
 5  172.16.0.18 (172.16.0.18)  0.520 ms    0.549 ms    0.517 ms
```

7.5   Practice: OSPF configuration

This final section describes how Quagga can be used to enable Linux-based OSPF routing in the topology of Fig. 7.9. Owing to the small size of the test network, only the backbone area is defined, although this configuration could be applied to any routing area. Assuming that zebra is already running, that all the interfaces have been configured with the proper IP address and that no other routing protocol daemon is active on any router, the ospfd daemon can be started and its command-line interface accessed via telnet using TCP port 2604:

```
[root@RouterA ~]# /etc/init.d/ospfd start
Starting ospfd:                                    [  OK  ]
[root@RouterA ~]# telnet 127.0.0.1 2604
Trying 127.0.0.1...
Connected to RouterA (127.0.0.1).
Escape character is '^]'.

Hello, this is Quagga (version 0.98.6).
Copyright 1996-2005 Kunihiro Ishiguro, et al.

User Access Verification

Password: ********
RouterA-ospfd> enable

Password: ********
RouterA-ospfd#
```

The OSPF operations require that each router is identified by a unique *router-id* in the form of an IP address. The router-id represents critical information for correct link state database creation and update and must be kept as stable as possible. Since this is a required parameter, when the OSPF routing process is activated a router sets its own router-id to one of the addresses assigned to its interfaces. However, since a physical network interface is subject to changes in time – it could be shut down or its address may be modified – the best practice used to ensure protocol stability is to assign a fixed IP address to the loopback interface, which is always active. The OSPF routers give priority to loopback interfaces when they automatically select the router-id. Alternatively, it is possible to set the router-id manually to a fixed and stable value, maintaining its uniqueness.

The command sequence to enable the OSPF routing process on router *A* and set a fixed router-id chosen among the IP addresses assigned to *A* (e.g., 10.0.0.9) is as follows:

```
RouterA-ospfd# configure terminal
RouterA-ospfd(config)# router ospf
RouterA-ospfd(config-router)# router-id 10.0.0.9
RouterA-ospfd(config-router)# exit
RouterA-ospfd(config)#
```

The OSPF protocol is capable of understanding the type of networks to which it is connected, since it must use this information to establish adjacencies in the most proper way. This means that when an Ethernet interface is detected, it is automatically classified as connected to a broadcast network and, as soon as OSPF is enabled on that interface, the DR election process begins. The test network of Fig. 7.9 has been built using Ethernet cross-over cables to realize point-to-point connections between Linux boxes with multiple network interfaces. As a consequence, all the links between routers are considered as broadcast multiple-access networks and a DR is elected on each of them.

However, it is possible for the router administrator to override this setting and explicitly set the type of network on each interface. For instance, it is possible to force router A to consider link AB as a point-to-point link:

```
RouterA-ospfd(config)# interface eth0
RouterA-ospfd(config-if)# ospf network point-to-point
RouterA-ospfd(config-if)# exit
RouterA-ospfd(config)#
```

To maintain a consistent topology, the same setting must be applied to the corresponding interface of router B:

```
RouterB-ospfd(config)# interface eth0
RouterB-ospfd(config-if)# ospf network point-to-point
RouterB-ospfd(config-if)# exit
RouterB-ospfd(config)#
```

Another important OSPF parameter is the cost assigned to an interface. The OSPF protocol has its own mechanisms to assign default cost values to network interfaces, which are typically higher for interfaces with lower bit rates. However, the router administrator is allowed to modify such settings by changing both the overall default behavior and the single interface cost. Since the interfaces used in the test network are all fast Ethernet, all the initial cost values are the same and are set by default to 10. To make the topology more interesting, it is possible to set the cost of link EF to a higher value, e.g., 20. This must be done on router E,

```
RouterE-ospfd(config)# interface eth0
RouterE-ospfd(config-if)# ospf cost 20
RouterE-ospfd(config-if)# exit
RouterE-ospfd(config)#
```

as well as on router F

```
RouterF-ospfd(config)# interface eth0
RouterF-ospfd(config-if)# ospf cost 20
RouterF-ospfd(config-if)# exit
RouterF-ospfd(config)#
```

The cost change is applied to both routers connected to the common link because costs are assigned to the single interface, meaning that the same point-to-point link can be seen with different cost values from the two routers. This is typical of link state protocols that work on a directed graph representation of the network topology. However, in the example considered here it has been preferred to maintain the same cost on each direction of link *EF*.

The final step consists of enabling OSPF on each network prefix that must be advertised by the protocol, specifying the routing area in which each prefix belongs. For instance, in the case of router *A*:

```
RouterA-ospfd(config)# router ospf
RouterA-ospfd(config-router)# network 10.0.0.8/30 area 0
RouterA-ospfd(config-router)# network 10.1.1.8/30 area 0
RouterA-ospfd(config-router)# network 192.168.0.0/24 area 0
RouterA-ospfd(config-router)# exit
RouterA-ospfd(config)# exit
RouterA-ospfd#
```

After all the routers have been properly configured, the link state packet exchange begins and each node is eventually able to create the complete network topology image and apply the Djkstra algorithm.

Once the link state protocol has converged, some interesting information can be extracted from each router about the OSPF process. For instance, this is what can be obtained on router *A*, including information about area zero and the size of the link state database:

```
RouterA-ospfd# show ip ospf
 OSPF Routing Process, Router ID: 10.0.0.9
 Supports only single TOS (TOS0) routes
 This implementation conforms to RFC2328
 RFC1583Compatibility flag is disabled
 OpaqueCapability flag is disabled
 SPF schedule delay 1 secs, Hold time between two SPFs 1 secs
 Refresh timer 10 secs
 Number of external LSA 0. Checksum Sum 0x00000000
 Number of opaque AS LSA 0. Checksum Sum 0x00000000
 Number of areas attached to this router: 1

 Area ID: 0.0.0.0 (Backbone)
   Number of interfaces in this area: Total: 3, Active: 3
   Number of fully adjacent neighbors in this area: 2
   Area has no authentication
   SPF algorithm executed 21 times
   Number of LSA 11
   Number of router LSA 6. Checksum Sum 0x00026863
   Number of network LSA 5. Checksum Sum 0x0003795f
   Number of summary LSA 0. Checksum Sum 0x00000000
   Number of ASBR summary LSA 0. Checksum Sum 0x00000000
   Number of NSSA LSA 0. Checksum Sum 0x00000000
   Number of opaque link LSA 0. Checksum Sum 0x00000000
   Number of opaque area LSA 0. Checksum Sum 0x00000000
```

Interface-specific information can also be extracted, such as network type, cost value, timer and neighbor status:

```
RouterA-ospfd# show ip ospf interface
eth0 is up
  Internet Address 10.0.0.9/30, Broadcast 10.0.0.11, Area 0.0.0.0
  Router ID 10.0.0.9, Network Type POINTOPOINT, Cost: 10
  Transmit Delay is 1 sec, State Point-To-Point, Priority 1
  No designated router on this network
  No backup designated router on this network
  Timer intervals configured, Hello 10, Dead 40, Wait 40,
    Retransmit 5, Hello due in 00:00:10
  Neighbor Count is 1, Adjacent neighbor count is 1
eth1 is up
  Internet Address 10.1.1.9/30, Broadcast 10.1.1.11, Area 0.0.0.0
  Router ID 10.0.0.9, Network Type BROADCAST, Cost: 10
  Transmit Delay is 1 sec, State DR, Priority 1
  Designated Router (ID) 10.0.0.9, Interface Address 10.1.1.9
  Backup Designated Router (ID) 10.1.1.10,
    Interface Address 10.1.1.10
  Timer intervals configured, Hello 10, Dead 40, Wait 40,
    Retransmit 5, Hello due in 00:00:06
  Neighbor Count is 1, Adjacent neighbor count is 1
eth2 is up
  Internet Address 192.168.0.254/24, Broadcast 192.168.0.255,
    Area 0.0.0.0
  Router ID 10.0.0.9, Network Type BROADCAST, Cost: 10
  Transmit Delay is 1 sec, State DR, Priority 1
  Designated Router (ID) 10.0.0.9, Interface Address 192.168.0.254
  No backup designated router on this network
  Timer intervals configured, Hello 10, Dead 40, Wait 40,
    Retransmit 5, Hello due in 00:00:01
  Neighbor Count is 0, Adjacent neighbor count is 0
lo is up
  OSPF not enabled on this interface
```

More detailed information about router A neighbors and adjacency status can be obtained with

```
RouterA-ospfd# show ip ospf neighbor

Neighbor ID    Pri   State          Dead Time   Address      Interface
10.0.0.17        1   Full/DROther   00:00:38    10.0.0.10    eth0:10.0.0.9
10.1.1.10        1   Full/Backup    00:00:34    10.1.1.10    eth1:10.1.1.9
```

Here the router-ids of B (10.0.0.17) and E (10.1.1.10) are listed, as well as the actual IP address used on the interface connected to A. Router A has a full adjacency with both of them and E is BDR on link AE. The lifetime of neighbor entries is also shown.

Owing to the cost increment made on link *EF*, path *ABCD* is now shorter than *AEFD*, even though it has the same hop count. This can be seen directly from *A*'s OSPF routing table:

```
RouterA-ospfd# show ip ospf route
============ OSPF network routing table ============
N    10.0.0.8/30              [10] area: 0.0.0.0
                              directly attached to eth0
N    10.0.0.12/30             [30] area: 0.0.0.0
                              via 10.0.0.10, eth0
N    10.0.0.16/30             [20] area: 0.0.0.0
                              via 10.0.0.10, eth0
N    10.1.1.8/30              [10] area: 0.0.0.0
                              directly attached to eth1
N    10.1.1.12/30             [40] area: 0.0.0.0
                              via 10.1.1.10, eth1
                              via 10.0.0.10, eth0
N    10.1.1.16/30             [30] area: 0.0.0.0
                              via 10.1.1.10, eth1
N    172.16.0.0/24            [40] area: 0.0.0.0
                              via 10.0.0.10, eth0
N    192.168.0.0/24           [10] area: 0.0.0.0
                              directly attached to eth2
```

The shortest path is then chosen when a traceroute is launched between the two hosts:

```
[root@host4 ~]# traceroute -n 172.16.0.18
traceroute to 172.16.0.18 (172.16.0.18), 30 hops max, 40 byte
 1   192.168.0.254   0.084 ms    0.085 ms    0.092 ms
 2   10.0.0.10   0.194 ms    0.130 ms    0.157 ms
 3   10.0.0.18   0.285 ms    0.257 ms    0.242 ms
 4   10.0.0.14   0.345 ms    0.312 ms    0.312 ms
 5   172.16.0.18   0.411 ms    0.399 ms    0.397 ms
[root@host4 ~]#
```

A summary of the link state database stored by *A* can be visualized with

```
RouterA-ospfd# show ip ospf database

        OSPF Router with ID (10.0.0.9)


            Router Link States (Area 0.0.0.0)
    Link ID     ADV Router     Age  Seq#        CkSum  Link count
    10.0.0.9    10.0.0.9       442  0x80000007 0x748c 4
    10.0.0.13   10.0.0.13      665  0x80000006 0x1b7d 2
    10.0.0.17   10.0.0.17      769  0x80000007 0x3639 3
    10.1.1.10   10.1.1.10      440  0x80000004 0x4b46 2
    10.1.1.14   10.1.1.14      389  0x80000007 0xc4f7 3
    10.1.1.18   10.1.1.18      385  0x80000006 0x91e4 2
```

```
                 Net Link States (Area 0.0.0.0)
Link ID      ADV Router      Age  Seq#         CkSum
10.0.0.13    10.0.0.13       665  0x80000001  0xd627
10.0.0.17    10.0.0.17       768  0x80000002  0x955e
10.1.1.9     10.0.0.9        442  0x80000001  0x9f6c
10.1.1.14    10.1.1.14       389  0x80000001  0xe30d
10.1.1.18    10.1.1.18      1330  0x80000001  0x8b61
```

The database consists of six router-LSAs and five network-LSAs. In fact, there are six routers in the test topology, so the number of router-LSAs must also be six. Then, of the six networks interconnecting routers, one has been explicitly configured as a point-to-point link, so it does not have a DR, whereas the other five links are considered to be broadcast multi-access networks and are advertised by the respective DRs, which generate the five network-LSAs.

It is interesting to have a closer look at some of the LSAs in the database, which are also exchanged through link state update packets. For instance, the link state originated by router *F*, which has router-id 10.1.1.18, can be extracted from *A*'s database with

```
RouterA-ospfd# show ip ospf database router 10.1.1.18

        OSPF Router with ID (10.0.0.9)

              Router Link States (Area 0.0.0.0)
  LS age: 537
  Options: 0x2  : *|-|-|-|-|-|E|*
  LS Flags: 0x6
  Flags: 0x0
  LS Type: router-LSA
  Link State ID: 10.1.1.18
  Advertising Router: 10.1.1.18
  LS Seq Number: 80000009
  Checksum: 0x8be7
  Length: 48
   Number of Links: 2

    Link connected to: a Transit Network
     (Link ID) Designated Router address: 10.1.1.18
     (Link Data) Router Interface address: 10.1.1.18
      Number of TOS metrics: 0
       TOS 0 Metric: 20

    Link connected to: a Transit Network
     (Link ID) Designated Router address: 10.1.1.14
     (Link Data) Router Interface address: 10.1.1.13
      Number of TOS metrics: 0
       TOS 0 Metric: 10
```

The LSA content starts with summary information, including age, type, advertising router and sequence number. This information is used to check whether the link state database is aligned between adjacent routers. Then the actual description of the router links follows, showing the characteristics of *EF* and *FD* as seen by *F*. Both are *transit networks*, i.e., packets routed to one of these networks are not necessarily directed to a destination address on the same network; in other words, routing of transit traffic is allowed on these links. Router *F* is the DR on link *EF* but not on link *FD*. Finally, the link cost is 20 for *EF* and 10 for *FD*.

An example of the network-LSA is given by the one advertised by *F* as acting as DR on link *EF*, which can also be extracted from *A*'s database:

```
RouterA-ospfd# show ip ospf database network 10.1.1.18

        OSPF Router with ID (10.0.0.9)

                Net Link States (Area 0.0.0.0)
  LS age: 1588
  Options: 0x2   : *|-|-|-|-|-|-|E|*
  LS Flags: 0x6
  LS Type: network-LSA
  Link State ID: 10.1.1.18 (address of Designated Router)
  Advertising Router: 10.1.1.18
  LS Seq Number: 80000004
  Checksum: 0x8564
  Length: 32
  Network Mask: /30
        Attached Router: 10.1.1.10
        Attached Router: 10.1.1.18
```

As expected, the network-LSA provides such information as the netmask used and the list of connected OSPF routers, *E* and *F* in this case. The information about the netmask is required, since it must be associated to the router interface address information included in the DR's router-LSA (shown above) to learn the complete network prefix of the multi-access network to be included in the routing table.

The last example of LSA is again a router-LSA, originated by router *D*:

```
RouterA-ospfd# show ip ospf database router 10.1.1.14

        OSPF Router with ID (10.0.0.9)

                Router Link States (Area 0.0.0.0)

  LS age: 1159
  Options: 0x2   : *|-|-|-|-|-|-|E|*
  LS Flags: 0x6
  Flags: 0x0
  LS Type: router-LSA
  Link State ID: 10.1.1.14
  Advertising Router: 10.1.1.14
  LS Seq Number: 800000cf
  Checksum: 0xda34
  Length: 60
   Number of Links: 3
```

```
Link connected to: a Transit Network
 (Link ID) Designated Router address: 10.0.0.13
 (Link Data) Router Interface address: 10.0.0.14
  Number of TOS metrics: 0
   TOS 0 Metric: 10

Link connected to: a Transit Network
 (Link ID) Designated Router address: 10.1.1.14
 (Link Data) Router Interface address: 10.1.1.14
  Number of TOS metrics: 0
   TOS 0 Metric: 10

Link connected to: Stub Network
 (Link ID) Net: 172.16.0.0
 (Link Data) Network Mask: 255.255.255.0
  Number of TOS metrics: 0
   TOS 0 Metric: 10
```

The interesting information here is the advertisement of the *stub network* 172.16.0.0/24, to which *D* is directly attached. A stub network is the opposite of a transit network, meaning that the only kind of traffic routed to this network is made of packets directed to a destination address included in the network prefix. Being the only router connected to this stub network, *D* is responsible for advertising the corresponding prefix.

To test the reconfiguration capability of OSPF, a ping session is started between the two hosts. During this session the link *CD* is forced to go down, causing OSPF to resynchronize the link state database on each router and recompute the shortest path. Looking at the resulting ping trace,

```
[root@host4 ~]# ping 172.16.0.18
PING 172.16.0.18 (172.16.0.18) 56(84) bytes of data.
64 bytes from 172.16.0.18: icmp_seq=1 ttl=60 time=0.497 ms
64 bytes from 172.16.0.18: icmp_seq=2 ttl=60 time=0.465 ms
64 bytes from 172.16.0.18: icmp_seq=3 ttl=60 time=0.451 ms
64 bytes from 172.16.0.18: icmp_seq=4 ttl=60 time=0.435 ms
64 bytes from 172.16.0.18: icmp_seq=5 ttl=60 time=0.441 ms
64 bytes from 172.16.0.18: icmp_seq=9 ttl=60 time=0.632 ms
64 bytes from 172.16.0.18: icmp_seq=10 ttl=60 time=0.554 ms
64 bytes from 172.16.0.18: icmp_seq=11 ttl=60 time=0.570 ms
64 bytes from 172.16.0.18: icmp_seq=12 ttl=60 time=0.565 ms
64 bytes from 172.16.0.18: icmp_seq=13 ttl=60 time=0.528 ms

--- 172.16.0.18 ping statistics ---
13 packets transmitted, 10 received, 23% packet loss, time 12000ms
rtt min/avg/max/mdev = 0.435/0.513/0.632/0.069 ms
```

it is apparent that OSPF is able to restore connectivity very quickly, less than four seconds in the case under observation. This is drastically shorter than the time required by RIP to deal with the same situation, demonstrating how much quicker link state protocols converge with respect to distance vector. Of course, repeating the traceroute now gives the alternative path:

```
[root@host4 ~]# traceroute -n 172.16.0.18
traceroute to 172.16.0.18 (172.16.0.18), 30 hops max, 40 byte
 1  192.168.0.254  0.092 ms    0.086 ms    0.092 ms
 2  10.1.1.10   0.213 ms     0.180 ms    0.175 ms
 3  10.1.1.18   0.322 ms     0.366 ms    0.352 ms
 4  10.1.1.14   0.398 ms     0.423 ms    0.379 ms
 5  172.16.0.18  0.533 ms    0.541 ms    0.507 ms
```

Recent versions of the Linux kernel allow Quagga to implement OSPF load balancing successfully when multiple shortest paths are found in the network. To test this functionality, the cost of link *EF* is set back to 10, so that both paths *ABCD* and *AEFD* now have the same cost. The OSPF routing table of router *A* now appears as follows:

```
RouterA-ospfd# show ip ospf route
============ OSPF network routing table ============
N     10.0.0.8/30          [10] area: 0.0.0.0
                           directly attached to eth0
N     10.0.0.12/30         [30] area: 0.0.0.0
                           via 10.0.0.10, eth0
N     10.0.0.16/30         [20] area: 0.0.0.0
                           via 10.0.0.10, eth0
N     10.1.1.8/30          [10] area: 0.0.0.0
                           directly attached to eth1
N     10.1.1.12/30         [30] area: 0.0.0.0
                           via 10.1.1.10, eth1
N     10.1.1.16/30         [20] area: 0.0.0.0
                           via 10.1.1.10, eth1
N     172.16.0.0/24        [40] area: 0.0.0.0
                           via 10.0.0.10, eth0
                           via 10.1.1.10, eth1
N     192.168.0.0/24       [10] area: 0.0.0.0
                           directly attached to eth2
```

Network prefix 172.16.0.0/24 is now reachable through two different paths with the same cost (40). The same information is written by `zebra` on the kernel routing table, although the `route` command is not capable of showing it. Therefore, the `ip route` command must be used to see it:

```
[root@RouterA ~]# ip route
10.1.1.12/30 via 10.1.1.10 dev eth1  proto zebra  metric 30 equalize
10.1.1.8/30 dev eth1  proto kernel  scope link  src 10.1.1.9
10.0.0.12/30 via 10.0.0.10 dev eth0  proto zebra  metric 30 equalize
10.0.0.8/30 dev eth0  proto kernel  scope link  src 10.0.0.9
10.0.0.16/30 via 10.0.0.10 dev eth0  proto zebra  metric 20 equalize
10.1.1.16/30 via 10.1.1.10 dev eth1  proto zebra  metric 20 equalize
172.16.0.0/24  proto zebra  metric 40 equalize
        nexthop via 10.0.0.10  dev eth0 weight 1
        nexthop via 10.1.1.10  dev eth1 weight 1
192.168.0.0/24 dev eth2  proto kernel  scope link  src 192.168.0.254
```

Router *A*'s kernel is natively configured to implement *per-flow load balancing*, meaning that different flows are equally routed along the two equal-cost paths available. However, to maintain packet sequence at the single flow level, packets belonging to the same flow are always routed on the same path. This is different from the case of *per-packet load balancing*, where the two paths are selected alternatively packet by packet, disregarding any flow information.

To verify how router *A* performs per-flow load balancing, it is necessary to generate two different flows between the two LANs. One flow is already known to the router, i.e., the ping session from the leftmost host to the rightmost one, which is started again. To generate a different flow, an additional ping session is started from another host on the LAN attached to router *A* and is directed to the same target host. Therefore, the two flows are both made of ICMP packets, both directed to 172.16.0.18, but one is generated from 192.168.0.4, the other from 192.168.0.11.

Using the command-line tool `tcpdump` to capture ICMP packets on interface `eth0` of routers *B* and *E* shows that the two flows are actually load balanced:

```
[root@RouterB ~]# tcpdump -n -i eth0 icmp
listening on eth0, link-type EN10MB (Ethernet), capture size 96 bytes
13:19:06.190250 IP 192.168.0.4 > 172.16.0.18: ICMP echo request, id 17778, seq 1
13:19:06.190525 IP 172.16.0.18 > 192.168.0.4: ICMP echo reply, id 17778, seq 1
13:19:07.169655 IP 172.16.0.18 > 192.168.0.11: ICMP echo reply, id 50801, seq 1
13:19:07.189222 IP 192.168.0.4 > 172.16.0.18: ICMP echo request, id 17778, seq 2
13:19:07.189476 IP 172.16.0.18 > 192.168.0.4: ICMP echo reply, id 17778, seq 2
13:19:08.169916 IP 172.16.0.18 > 192.168.0.11: ICMP echo reply, id 50801, seq 2
13:19:08.188213 IP 192.168.0.4 > 172.16.0.18: ICMP echo request, id 17778, seq 3
13:19:08.188466 IP 172.16.0.18 > 192.168.0.4: ICMP echo reply, id 17778, seq 3
13:19:09.170214 IP 172.16.0.18 > 192.168.0.11: ICMP echo reply, id 50801, seq 3

[root@RouterE ~]# tcpdump -n -i eth0 icmp
listening on eth0, link-type EN10MB (Ethernet), capture size 96 bytes
13:19:07.171237 IP 192.168.0.11 > 172.16.0.18: ICMP echo request, id 50801, seq 1
13:19:08.171504 IP 192.168.0.11 > 172.16.0.18: ICMP echo request, id 50801, seq 2
13:19:09.171480 IP 192.168.0.11 > 172.16.0.18: ICMP echo request, id 50801, seq 3
```

These capture traces also show that the replies from the target host are always routed by *D* on the same path, i.e., *DCBA*. Elaboration on the possible reasons for this behavior is left to the reader.

8 Wide area networks and user access

This final chapter is dedicated to a brief overview of some of the most important solutions used by network operators, on the one hand, for providing access to their customers, on the other, as wide area network (WAN) technologies are adopted in the network core. The chapter starts with an overview of the xDSL family, one of the most popular access technologies adopted today. Then a historical perspective of WAN technologies is given, going from the X.25 and ISDN solutions to the more recent frame relay service and ATM protocol. The attempt made by IETF to join the flexibility of the connectionless IP world with the advantages of the connection-oriented ATM approach is, finally, the basis for introducing the principles of MPLS, a very popular technology currently used by operators to offer enhanced IP-based connectivity services to their customers, including quality of service guarantees and tunneling capabilities. Given the importance of the last topic, the chapter closes with a section dedicated to a practical example of MPLS router configuration.

8.1 The xDSL family

The xDSL acronym refers to a set of solutions and technologies developed for supporting the transmission of high speed data in existing copper access networks. It also supports data transmissions at long distances but at lower bit rates.

The ITU-T recommendation G.995.1 [91] provides an overview of the digital subscriber line family. In the following ADSL, with the recent ADSL2 and ADSL2+, HDSL, SHDSL, VDSL and VDSL2 will be presented.

The ITU-T G.992.1 [92] recommendation, published in 1999, specifies the physical layer characteristics of the *asymmetric digital subscriber line* (ADSL) technique, which is employed on the communication link between the end user and the network access node, the so-called subscriber line. It allows for high speed access to wide area networks. It has been conceived and developed for exploiting the existing, already installed, copper twisted pairs that connect residential users to the public switched telephone network.

So far, these twisted pairs have been employed for carrying audio signals in the 0–4 kHz bandwidth but, actually, they are able to operate at frequencies of the order of up to 1 MHz. As a matter of fact, many network operators have been offering this technology for accessing the Internet at high speed for several years.

The ADSL is asymmetric, which means that the downstream communication channel capacity from the network node toward the user is greater than the upstream capacity in the opposite direction. This asymmetry is suitable for the way a typical user operates on the Internet, where the amount of data downloaded is greater than that uploaded. The ADSL has been developed for supporting a minimum bit rate of 6.144 Mbit/s in download and 640 kbit/s in upstream.

The basic idea of this solution is to separate the voice and the data services in the frequency domain: the lowest 25 kHz of bandwidth are reserved for circuit switched phone calls, with the 4–25 kHz guard band to prevent cross-talk between voice and data. Different techniques can be used for managing the upstream and downstream bandwidths. The first one is frequency division modulation, which allows parallel data transmissions within the up and downstream bandwidths. Another is echo suppression.

The ADSL employs the *discrete multi-tone* (DMT) modulation technique, where the data flow is transmitted in parallel over multiple carriers at different frequencies. The 1.1 MHz available bandwidth is decomposed into 256 sub-channels of roughly 4 kHz: channel 0 is for voice, channels one to five are not used but considered as guard band, one channel is for controlling the transmission and another for the reception, the remaining ones are for user data transmission and, usually, 32 of them are assigned to the upstream.

Before the transmission, the DMT modem tests the quality of the access link, in terms of signal-to-noise ratio, by sending probing signals. Then, user data flow is distributed over the carriers, not proportionally but as a function of the estimated quality. This means that the channels with better estimated quality are loaded with more data bits and each channel can have a transmission rate ranging between 0 and 60 kbit/s (Fig. 8.1).

The ADSL is typically employed at a distance of up to 5.5 km so that it can serve most residential area users, ranging from 75% to 95% of the customers in Western countries.

In 2005, ITU-T delivered two enhancements of ADSL. Recommendation ITU-T G.992.3 [93] describes ADSL2, which allows for data transmissions up to a minimum of 8 Mbit/s downstream and 800 kbit/s upstream, while ADSL2+, ITU-T G.992.5 [94], compared with ADSL2, employs double the downstream bandwidth. Five hundred and twelve subcarriers (channels) are employed up to the 2.208 MHz frequency. It supports data rates up to a minimum of 16 Mbits/s downstream and 800 kbit/s upstream. Of course, the requirements in terms of maximum distance for these enhancements drop to 1–2 km.

The *high bit rate digital subscriber line* (HDSL), described in ITU-T G.991.1 (1998) [95], is a two or three copper twisted pairs bidirectional and symmetrical transmission system with a bit rate of 2048 Mbit/s (E1) or 1544 Mbit/s (T1). It employs echo

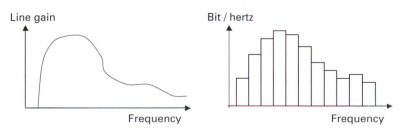

Fig. 8.1   Discrete multi-tone modulation

cancellation for implementing full-duplex, so that both directions of data transmission can be supported on one twisted pair. Three systems can actually be used, one with a 784 kbit/s bit rate over each twisted pair, a second with a 1168 kbit/s over two pairs only and a third with a 2320 kbit/s on one pair only. Two possible line codes can be used, pulse amplitude modulation, 2B1Q, and carrierless amplitude or phase modulation, CAP. However, since it requires two twisted pairs, it is not very suitable for serving residential users, who typically have just one.

The *single-pair high-speed digital subscriber line* (SHDSL), ITU-T G.991.2 (2003) [96], is a transmission system for full-duplex data transmission over a single twisted pair. Echo suppression is used for full-duplex communications. Symmetric data transmissions are supported at bit rates ranging from 192 to 2312 kbit/s using a trellis coded pulse amplitude modulation line code.

The *very high-speed digital subscriber line* (VDSL), ITU-T G.993.1 (2004) [97], is an asymmetric and symmetric transmission system over one twisted pair. It allows data rates up to tens of Mbit/s by using up to 12 MHz of bandwidth. Full-duplex is realized by means of frequency division multiplexing (FDM). Bandwidth above 138 kHz up to 12 MHz is divided into four bands, dedicated alternatively to downstream and upstream. The VDSL2, ITU-T G.993.2 (2006) [98], is an enhancement of VDSL, which supports asymmetric and symmetric full-duplex data communications at data rates up to 200 Mbit/s using a bandwidth up to 30 MHz. This solution seems very promising for delivering multimedia applications to residential users, even if the reduced maximum distance may represent a limit to their broad diffusion.

8.2   The X.25 network

The ITU-T X documents standardize data networks and open system communications. The X.25 network [99] was one of the first packet-switched data network solutions and takes into account the first three lower layers of the OSI model. As the standard states, X.25 specifies the protocol to be used at the "interface between *data terminal equipment* (DTE) and *data circuit equipment* (DCE) for terminals operating in packet mode and connected to public data networks by a dedicated circuit." This recommendation was revised in 1993 to 1996 and approved in October 1996. The access bit rates at the interface can be up to 2 Mbit/s.

As far as layer-1 is concerned, X.25 recalls the X.21 document [100], for synchronous point-to-point lines on public data networks, the X.21bis [101], for interfacing DTEs to synchronous V-series modems, and the X.32 [102] for accessing packet data networks through PSTN. The V set of standards describes data communications over telephone lines and the documents V.36 [103], V.37 [104] and V.38 [105] regard wideband modem interfaces. As for layer-2, X.25 provides a detailed description of the LAP-B (link access procedure – balance mode) protocol, which is a link access procedure belonging to the HDLC family. Layer-3 reports the packet format, which will be placed within the information field of a HDLC frame.

Generally speaking, X.25 was conceived for virtual circuits, which can be either permanent (*PVC: permanent virtual circuit*) or switched (*SVC: switched virtual circuit*). Permanent virtual circuits are suitable for frequent and long connections toward a predefined destination. Switched virtual circuits, on the other hand, are typically suitable for communications toward different destinations and, later, datagram transfer mode was also added.

Packets are represented as pages with rows of eight bits each. The first three bytes of each are quite similar. The first one contains a four-bit field, which represents the group number, and a second one, called the GFI, or general format identifier. The second byte has the channel number and the third specifies the packet type. Sequential bytes may be present and they depend on the packet type.

Since X.25 is a packet-based connection oriented protocol, the first operation a terminal has to perform to access the network is to open a virtual channel by sending an open request packet. This virtual channel is identified by a field composed by two parts: a four-bit group number, which leads to 16 possible groups, and an eight-bit channel number, which means up to 256 channels per group. In summary, 4096 virtual channels can be addressed on an interface.

To avoid a collision during the connection set-up between user terminal and network for the number assignment, it has been agreed that the former picks the lowest available number while the latter chooses the highest available one.

The service of virtual channel or connection establishment is *acknowledged*, so that the called terminal has to decide whether to accept the incoming connection or not. The calling station sends a control packet of the type *call-request* asking to open a connection. The network informs the destination station of this request through another control packet, *incoming-call*. If it concerts to open the connection, it replies to the network with a *call accepted* packet. Finally the network informs the caller, with a *call connected* packet, that the connection has been established and is operating.

The one-byte field *type* specifies whether the packet carries user data or control information. The first bit (from right to left) is equal to 0 for data packets and to 1 for control packets.

There are five packet types, namely, *call set-up and clearing*, *data and interrupt*, *flow control and reset*, *restart* and *diagnostic*.

The *call request* packet format has the type field equal to 00001011 and belongs to the first type of packet. To get to the destination, some information is required, which means that an address must be specified. After the type field, up to 15 digits can be reported but, since the corresponding number may be of variable length, it is necessary to have before it a byte, composed of two nibbles, which represents, in binary digits, the number of digits of the sender and destination address; thus, with four bits, at most, there are 15 digits. In X.25, actually, the address field for the caller is not mandatory. Every digit of the address is coded following BCD (four bits per digit) and if the sum of the two address digits is odd a group 0000 is added to the last word. Overall, these addresses occupy at most 16 bytes. A field of *facilities* comes next, whose first byte has two bits at 0 and the other six bits representing the length. Having two zeros, this field can never exceed 64 bytes ($2^6 - 1$). It can contain features and characteristics of the connection requested to the network, such as, for instance, cost charging to the destination.

The *call accepted* packet format is similar to the *call request* one but in responses it is necessary to invert caller and callee addresses. The facility field contains the response to the facilities previously required.

Only these packets include the complete addresses because, once the connection is established, all other packets can refer to that connection by means of the channel and group number.

Other packets are used to close a virtual circuit, such as *clear request*, to which the network replies with a *clear confirmation*.

The data packet always reports the group and channel number and the first bit of the type field is 0.

While in control packets the remaining bits of the type field are for specifying the different controls, in data packets they are divided into two groups, $P(R)$ and $P(S)$, plus one bit, the M bit. $P(R)$ and $P(S)$ are very similar to $N(R)$ and $N(S)$, used by the HDLC data-link layer protocol, and are used for end-to-end flow control through a sliding window mechanism.

$P(R)$ reports the number of the next expected packet on that connection, so that through $P(R)$ the acknowledgment is at packet level. It is worthwhile reminding ourselves that layer-3 does not perform any error recovery since this is a layer-2 task. If, on the other hand, layer-2 fails, layer-3 can detect the event of out-of-sequence packets but it simply signals it to upper layers and performs the *reset* operation, without any attempt to recover the error. Also, the data packet is inserted within a layer-2 frame and on a link all frames are checked by layer-2 algorithms so that it may seem redundant to use the fields $N(R)$ and $N(S)$ in layer-2 protocols. Actually, at layer-3 the task is different since through $P(R)$ and $P(S)$ the control is made for every virtual channel.

The data field has a limited length, for instance 128 bytes, and since layer-3 data come from layer-4, some fragmentation may be required. Long files are fragmented in many parts and the presence of following fragments is notified by setting the M (*more*) bit to 1. The last part of a sequence has $M = 0$.

Additional bits are used within a data packet header. The Q (*qualifier*) bit is not controlled by X.25 but by upper layers, which can exchange controls without messing with layer-3 data. In layer-3 data packets, $Q = 0$. The *modulus* bits (bits 6 and 5) can be 01 or 10 only. With 01, data packets have normal format while with 10, packets have an extended format. The extended format packet doubles the type field and seven bits are given both to $P(S)$ and to $P(R)$, so that transmission windows can be as long as 128. Layer-3 windows larger than layer-2 ones are justified by network delays longer than link-level delays. This allows for a more effective use of network resources.

Interrupt packets are urgent packets, which overtake all data packets and never queue behind them. This is a way to deliver urgent and critical data, such as alarms, warnings, and so on. Their correct reception has to be acknowledged through a proper packet (*interrupt confirmation*), which the receiver has to send back as soon as possible. In conclusion, these packets realize a sort of *stop-and-wait* low-speed communication channel in parallel to the established virtual channel.

A third type includes the packets for performing flow control and packet retransmissions. For instance, the *reject* packet aims to retransmit packets of the data flow starting from the one reported in $P(R)$. The *reset* packet operates on single virtual channels and determines the reset of all state variables and the flush of all current packets in the channel; in other words, $P(R)$ and $P(S)$ are reset.

The fourth type, *restart* packets, involve the whole interface, which means that all active virtual channels on that interface are terminated and must be set up again. Actually, this is strictly true for dynamic, switched virtual channels because permanent virtual channels are unaffected.

Diagnostic packets, not necessarily present on every network, have one or more bytes for additional information about the reason for, for instance, a restart or a reset.

If a little amount of data has to be transmitted so that the establishment of a full connection might not be worthy, X.25 has defined the *fast select* facility. A call request packet can be sent with the fast select facility only to realize a sort of connectionless or datagram communication.

The data field can be as long as 128 bytes. It is worth emphasizing that since no connection is established, this datagram overhead is bigger than a regular data packet, since here the complete destination and sender addresses have to be included.

8.3   Integrated services digital network

The *integrated services digital network* (ISDN) [106], is a digital wide area telecommunications network standardized by ITU-T in the I series documents. One of its main goals is to provide an infrastructure suitable for supporting data and voice applications at the same time, by means of a limited number of user-network interfaces. It allows the support of circuit and packet switched connections at multiples of 64 kbit/s.

The definition of ISDN started in the 1980s through the I series documents. These documents are grouped into seven sets, from I.100 to I.700. Set I.100 reports the general structure, I.200 the service capabilities, I.300 the overall network aspects and functions, I.400 the user-network interfaces, I.500 the internetwork interfaces, I.600 the maintenance functions and I.700 the equipments for the broadband-ISDN.

Users are connected to an ISDN switch with a digital transmission line suitable to carry different communication channels, namely B, D and H channels.

The B channel is at 64 kbit/s and it is the basic channel for user access. It can be employed both for voice and for data communications. It can be used for establishing different connections, namely, circuit-switched, packet-switched through X.25, frame mode with circuit to a frame relay node and data exchange with LAP-F, and, last, semi-permanent, established by the network provider, which is just like a dedicated channel.

Table 8.1. Characteristics of ISDN channels

B channel (64 kbit/s)	D channel (16 kbit/s)
Digital voice	Signaling
64 kbit/s (PCM)	Basic
32 kbit/s	Extended
High-speeed data	**Low-speed data**
Packet switching	Videotex
Circuit switching	Teletex/terminali
Others	**Telemetry**
Fax	Emergency and alerts
Slow motion video	

The D channel is usually employed to carry signaling information related to circuit switching of the associated B channels and, sometimes, to carry low-speed packet data (100 bit/s) when there is no signaling [107].

The H channels are installed for high-speed network accesses (384 (H0), 1536 (H11), 1920 (H12) kbit/s) and can be employed to support advanced applications, such as high-speed fax, video, high-speed data, high-quality audio and video-conferences.

The B and D channels are actually grouped into two main structures, provided and managed by the network provider, as shown in Table 8.1. The former is the basic access, given by two full-duplex B channels and by a full-duplex D channel at 16 kbit/s. This is mainly used to provide access to residential areas and for SOHO and it allows for simultaneous data and voice communication. The frame overhead due to proper synchronization takes the bit rate from 144 to 192 kbit/s. The latter is the primary access used in networks with higher speed requirements. In Europe, it is at 2.048 Mbit/s and it is given by 30 B channels at 64 kbit/s plus one D channel at 64 kbit/s. In the USA, Canada and Japan this access is at 1.544 Mbit/s with 23 B channels and one D channel at 64 kbit/s. Also, a flexible access $nB + D$, with $1 < n < 30$, can be configured, depending on user needs.

Network access with H channels can be twofold: the first is implemented with H0 channels at 384 kbit/s; either with three of them plus one D channel at 64 kbit/s or with four of them: alternatively, five of them can be used plus a D channel at 64 kbit/s to get the 2.048 Mbit/s rate. The latter is realized by means of H1 channels: the H11 channel is at 1536 kbit/s for the E1 interface, whereas H12 is at 1920 kbit/s plus a D channel at 64 kbit/s for the E1 interface.

The ITU-T, in the document I.324 entitled *ISDN Network Architecture* has defined *functional groupings* and *reference points* for user access (Fig. 8.2). Functional groupings are a finite set of possible configurations and functions of the ISDN arrangements and devices. The reference point is a conceptual point between two functional groupings: it may correspond to the physical interface between pieces of equipment.

As regards functional groupings, four of them have been defined. The NT1 (network termination 1) has functions related to the physical termination (layer-1 ISO/OSI) of ISDN at the user sit. It is managed by the network provider and performs, for instance,

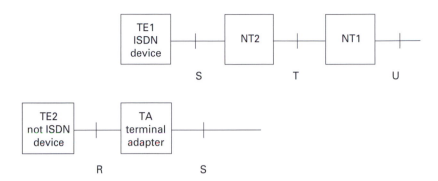

Fig. 8.2   Functional groupings and reference points

synchronous time division multiplexing to support the basic access. The NT2 performs functions up to layer-3, e.g., digital PBX or LAN. The TE1 (terminal equipment type 1) represents a device that supports the ISDN interface, such as a digital telephone and fax. The TE2, on the other hand, represents non-ISDN devices, with, for instance, the TIA or EIA-232-F interface, which need terminal adapters.

As regards the reference points, three of them have been defined. T (terminal) corresponds to the ISDN network termination at the user site and it is the border between provider and user devices. S (system) represents the interface between the user ISDN terminal and the network device. R (rate) is the non-ISDN interface for non-ISDN devices, for instance toward TIA/EIA-232-F.

The ISDN can provide four types of service for end-to-end communications. Circuit switching over the B channel is realized by means of NT1 or NT2 (layer-1 functions). The D channel is also employed for signaling between user and network, to establish and to tear down the circuit. For instance, the D channel may be used to carry signaling messages of SS7 (signaling system number 7). Semi-permanent connections can be established over the B channel between predefined user sites. The network interface provides layer-1 functions only, without any signaling support, since everything is managed by the network provider. Packet switching is supported and provided over the B channel and the user operates as the DTE while the network node is seen as the DCE. Also, the connection between the user and network nodes may be either permanent or circuit switched and, for the latter, D channel is used for managing the circuit.

8.4   The frame relay service

Frame relay is a standard defined by the ITU-T, which has some similarities with X.25, since it refers to the DCE–DTE interface allowing for the multiplexing of different virtual circuits over the same link. On the other hand, there are a lot of differences because frame relay is basically a layer-2 standard so that it does not have, as X.25, a network layer. Frame relay has been conceived for reliable high-speed lines where the bit error rate is not as high as in X.25 lines, so that errors are not corrected on a link basis as in X.25 but the matter of dealing with errors is shifted to end nodes.

It is worth remembering that in the 1990s a strong deployment of optical fibers started, replacing copper lines, such as coaxial cables, for medium- to long-distance communications. One of the main features of optical fibers is the extremely low bit error rate compared with coaxial cables. This has made all link-by-link error checks useless, imposing a delay in the data forwarding for their computation. In essence, the network bottleneck has moved from transmission lines to nodes, which were suddenly loaded by high-speed lines that were sending in much more data to be switched.

Thus, frame relay has been the first solution toward a high-speed network in a wide area. It has been widely used for effectively connecting routers in geographical areas replacing X.25. Frame relay has been defined for T1/E1 speeds even if it may be extended to T3/E3.

The ITU-T standard Q.922 [108] specifies the frame structure and procedures of the data-link layer to support frame mode bearer services in the user plane, as reported in ITU-T I.233 [109]. This definition of layer-2 is based on the extension of ITU-T Q.921 [107] LAP-D (link access procedure on the D-channel), which becomes LAP-F (link access procedure to frame mode bearer services).

The LAP-F frame has a flag just like HDLC to define the beginning and the end of the frame itself. It employs the bit stuffing technique for transmission transparency. The LAP-F is, in its turn, composed of two parts, DL-CORE (data link core protocol), reported by ITU -T I.233, and DL-CONTROL (data link control protocol), which is the remaining part of the frame (Fig. 8.3). In detail, the DL-CORE fields are flag, address, which is of variable length (two, three or four bytes) with some congestion control functions, and FCS, frame check sequence with a cyclic redundancy check over two bytes.

It is worth describing the address field in more detail. It is composed of different subfields (Fig. 8.4): EA, extended address, is a bit, which when set to 1, means that this byte is the last of the address field; the C/R bit is for future use. The FECN (forward explicit congestion notification) bit is set to 1 by nodes to inform the next routers that the data path is congested. The BECN (backward explicit congestion notification) bit is set to 1 by nodes to inform routers that the data path in the opposite direction has some congested links. The DLCI (data-link connection identifier) is the virtual channel identifier; it is usually 10 bits long and it has a link-by-link meaning. The DE (discard eligibility indicator) bit means, when set to 1, that the current frame may be discarded in the event of network congestion. Finally, the D/C (DLCI or DL-CORE control indicator) bit indicates whether the remaining six bits have to be considered as DLCI or as DL-CORE control.

The frame relay service allows for the set-up of virtual circuits between DTE and frame relay. It establishes connection-oriented channels, which were permanent in the beginning, while the dynamic circuit set-up was introduced later on. Permanent virtual circuits (PVCs) are set up and configured by a network provider, so this service is mainly

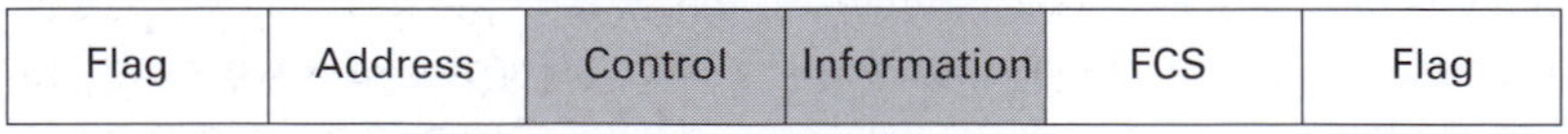

Flag	Address	Control	Information	FCS	Flag

Fig. 8.3   The frame relay frame format

Upper DLCI			C/R	EA0
Lower DLCI	FECN	BECN	DE	EA1

Upper DLCI			C/R	EA0
DLCI	FECN	BECN	DE	EA0
Lower DLCI or DL-CORE control			D/C	EA1

Upper DLCI			C/R	EA0
DLCI	FECN	BECN	DE	EA0
DLCI				EA0
Lower DLCI or DL-CORE control			D/C	EA1

Fig. 8.4   The address field

suitable for users who need a predetermined amount of bandwidth among well-defined network access points. Correspondingly, the cost is a function of the number of PVCs without considering the actual use, thus penalizing bursty sources.

The need for network flexibility has pushed for the definition of a switched virtual circuit (SVC). Users can now ask for a dynamic set up and release of virtual circuits through the employment of automatic procedures and this requires the definition of user-network and node-to-node signaling protocols. This on-demand service allows the network to provide connections within a few minutes in a cheap way, so as to give network accesses and not just edge-node interconnections.

Differently from the user plane, all nodes have also layer-3 functions for the SVC management.

The ITU-T Q.933 [110] specifies the signaling to establish, maintain and release virtual circuits; Q.933 control information is carried within LAP-F frames. In particular, Q.933 defines the access methods through an ISDN interface. The first option is a circuit-switched access to a remote frame handler (RFH) by means of B or H ISDN channels between a user and a RFH using the Q.931 signaling [111]. The second option is frame relay access to a frame handler (FH), usually placed within the local ISDN switch, where frames are packet switched.

A frame relay network can be realized with a set of frame relay switches: core nodes, which switch incoming frames by just looking at the DLCI or, more generally, by just processing the DL-CORE, whereas the edge nodes work both on the DL-CORE and on the DL-CONTROL fields (Fig. 8.5). This approach is called *core-edge* because several functions are performed edge-to-edge only, such as error recovery and flow control.

Frame relay has defined procedures for traffic control. Traffic entering the network is controlled following parameters agreed during connection set-up. Let us suppose that

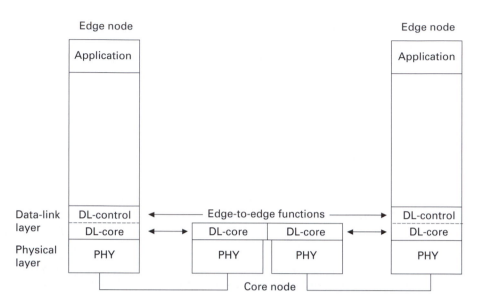

Fig. 8.5   Frame relay core-edge approach

traffic sources are variable bit rate (VBR). Thus, the quality of service provisioning from the network can be guaranteed statistically only.

Three main parameters are employed: the *committed rate measurement interval*, T_c, i.e., the time interval during which connection traffic parameters are evaluated; the *committed burst size*, B_c, which represents the maximum number of bits sent by the source in T_c which the network commits to accept; the *excess burst size* B_e, which is the maximum number of bits sent by the source in excess of B_c in T_c, which the network will do its best to forward but without any guarantee.

With these parameters, two averages can be defined, which represent the available capacity to users, with or without guarantee. The *committed information rate*, *CIR*, is the average access capacity expressed in bit/s to the frame relay network or, in other words, is the bit rate which the network commits to deliver successfully to the destination and it is defined as $\text{CIR} = B_c/T_c$. The *excess information rate*, *EIR*, is the access capacity expressed in bit/s in excess of CIR, which the network accepts anyway from the user without providing any guarantee of successful forwarding: these data can be discarded any time, anywhere in the network in the event of congestion because these frames are marked by access nodes by setting the bit $DE = 1$. It is straightforward to see that $\text{EIR} = B_e/T_c$.

The ITU-T Recommendation I.370 [112] describes the congestion management strategies for frame relay, intended to work for access bit rates up to 2048 kbit/s. It is worth remembering that frame relay is a simplified network compared with X.25, in which layer-2 mechanisms such as sliding windows are not possible, since inter-node flow control is not present anymore. Congestion control is then performed in cooperation between network and end users: the network monitors its state while end users will limit the traffic injected into the network. Three main approaches can be adopted. The first is frame discarding, which is based on the DE bit. The second is the explicit congestion notification: to prevent congestion, every node controls its buffer occupancy level and, once a given threshold is crossed, a congestion signaling procedure is activated to all users with active connections by means of the FECN and BECN bits. The third one is the implicit congestion notification, which means that the end user realizes congestion due to frame losses, by means of LAP-F or, at an upper layer, TCP.

8.5   B-ISDN and ATM

In 1990, the former CCITT approved some recommendations on the *broadband ISDN* (B-ISDN) in the I series. Among them, I.113 is a vocabulary of terms for broadband aspects of ISDN and I.121 reports the broadband aspects of B-ISDN. Recommendation I.113, revised in 1997, defines broadband as a service or system requiring transmission channels that can support bit rates greater than the primary rate. The term B-ISDN is used to emphasize the broadband aspects of ISDN with the goal of having a more general definition of ISDN, which can include all kinds of service, broad and narrowband.

The ITU-T has declared the *asynchronous transfer mode* (ATM) as the transfer mode for implementing the B-ISDN. Recommendation I.326 (2003) [113] defines the functional architecture of transport networks based on ATM.

It is worth mentioning the ATM Forum as an important organization, mainly of vendors and operators, constituted in 1991 to speed up the developing process of ATM devices and services, mainly toward the private field. It has contributed to the study and definition of solutions for the issues of congestion control, traffic management, new applications and adaptation layers. In 2004, the ATM Forum merged into the MPLS and Frame Relay Alliance, originating the MFA Forum.

The ATM is a packet-oriented transfer mode. It has a labeled multiplexing scheme with a slotted transport capacity. The switching function adopts a connection-oriented routing and forwarding is based on store and forward with logical resource assignment and asynchronous functioning. Signaling and user data are carried over separate virtual channels and the multiplexed data flow is organized into fixed length units called cells. Thus, ATM is a packet-oriented transfer mode based on the asynchronous time division multiplexing of fixed length cells. These cells are transparently transferred through the network and no error check is performed within the network. The ATM can carry any type of service and, to this end, it defines different adaptation layers to meet the different service needs. Routing is connection oriented and it is performed by means of virtual channels and virtual paths, where a virtual path can be considered as a virtual channel aggregation. Within a virtual channel, the cell sequence is guaranteed.

Network resources can be assigned on a semi-permanent or per call basis and they may be re-negotiated during the call. These resources can be expressed in terms of bit rates or quality of service. For each virtual channel, a portion of bandwidth is assigned depending on the number of multiplexed connections, traffic burstiness and required QoS. The source peak bit rate is also an important parameter for managing the available bandwidth. The QoS of a connection is expressed as a function of cell loss, cell delay and

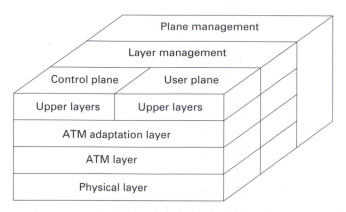

Fig. 8.6   The B-ISDN protocol reference model

jitter. Also, the cell header has a cell loss priority bit (CLP bit) for managing a two-level priority on a virtual channel in the case of congestion, discarding lowest priority cells first.

The ITU-T Recommendation I.321 [114] describes the B-ISDN protocol reference model (Fig. 8.6), considering, in particular, the ATM transfer mode. It is composed of three planes. The user plane manages user data flow. The control plane provides all the required functions for controlling the channels; it manages the signaling to establish, maintain and release the channels. The management plane is, in its turn, divided into two parts: a plane management, which has a non-layered structure and deals with the different plane management and coordination, and a layer management, which provides the functions for managing the resources of each layer and handles the operation and maintenance (OAM) information flows for a specific layer.

In the user plane, above the physical layer, there are two layers specific to ATM: the ATM layer, which provides the cell transfer for all services, and the ATM adaptation layer, which provides service-dependent functions to the upper layer.

The physical layer consists of two sublayers, the physical medium (PM) and the transmission convergence (TC). The physical medium deals with the function strictly according to the medium used, such as transmission and coding and bit timing. Transmission convergence allows for the ATM definition, completely independently of the physical medium used. It performs all the functions required to transform a flow of cells into a bit flow ready to be transmitted over the physical medium. To this end, many functions are achieved. *Transmission frame generation and recovery* is the function which performs the frame generation and recovery. *Transmission frame adaptation* is the function necessary to process the cell flow according to the payload structure of the transmission frame and to take the cells out of it. The function of *cell delineation* prepares the cell flow so that the receiver correctly recognizes a flow of cells with the proper boundaries. *Head error check sequence generation and cell header verification* is the function which performs, in the sending direction, the header error check (HEC) computation and its insertion in the header and, in the receiving direction, the cell header error checks. Finally, *cell rate decoupling* performs insertion or suppression of idle cells to adapt the rate of valid ATM cells to the transmission system payload.

The ATM layer performs all the functions related to the ATM cell header, it is physical-medium independent and common to all types of service. It performs cell multiplexing and de-multiplexing, which means that it combines cells from different virtual paths and virtual channels into a unique cell flow, and vice versa. It also performs cell-header generation and extraction. Finally, the ATM layer is responsible for label switching, one of the main features of ATM, which is the translation or re-writing of the virtual circuit identifiers in the core nodes.

The ITU-T Recommendation I.361 [115] describes the B-ISDN ATM layer specification, including the cell format. The cell is the ATM basic unit and consists of two fields, a five-byte header and a 48-byte information field. Two different formats are also defined, one for the user–network interface (UNI) and the other for the network–network interface (NNI). This structure is fixed length and this is very important because it is possible to determine network queueing models for computing cell delay and loss. Of course, this is particularly important for analyzing the performance of real-time applications. Fixed-length units are also very useful to design high-speed switching systems, since most of the processing can be done in hardware.

The UNI cell header (Fig. 8.7) has the first octet divided into two parts: four bits of generic flow control (GFC) and four bits as virtual path identifier. Generic flow control can be used for flow control in virtual channels and for managing short-term congestions. The VPI has four more bits in the second octet and then 16 bits of virtual channel identifier start. It is worth remembering that the couple VPI/VCI identifies a connection and allows for realizing a connection-oriented dynamic resource allocation without static bandwidth allocation. Then come three PTI (payload type identifier) bits and the octet is terminated by the CLP bit. The PTI specifies whether the cell contains user data or some signaling. The last octet of the header is occupied by HEC. The NNI cell header is

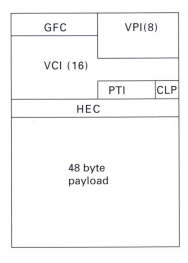
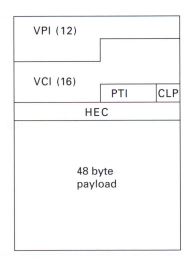

ATM UNI   ATM NNI

Fig. 8.7   ATM cell formats

different in the first octet only, since it does not have the GFC field, so that the VPI has now 12 bits.

The ITU-T I.363 Series [116] specifies the B-ISDN *ATM adaptation layer* (AAL). The AAL has the goal of increasing the service provided by the ATM layer to support service dependent requirements. The AAL supports multiple protocols to meet the needs of the different AAL service users. The AAL is composed of two sublayers, convergence sublayer (CS) and segmenting and reassembling (SAR). Different combinations of CS and SAR give different AAL services and if the ATM layer service is enough for some applications the AAL would be empty. An AAL-SDU (service data unit) can be carried from one AAL-SAP (service access point) to another through the ATM network. An AAL user can select the proper AAL-SAP to employ to get the required QoS.

The SAR sublayer functions are performed on an ATM-SDU basis. The SAR sublayer fragments the information flow from the upper layer to fit the ATM payload and, on the other hand, reassembles the payload of the received cells into higher-layer format.

The CS functions may include the blocking of user information to form 47 octet SAR-PDU payload, the handling of cell delay variation and the handling of lost and out-of-sequence cells. Also, it may be further divided into two sublayers, the common part convergence sublayer (CPCS) and the service-specific convergence sublayer (SSCS).

Four types of AAL have been developed for four different *classes of service*, defined as a function of three parameters: the time relationship between source and destination, the bit rate and the connection mode (Fig. 8.8). Actually, four out of the eight possible combinations have mainly been considered for defining classes of service.

Class A is for applications requiring a strict time relationship between source and destination, constant bit rate (CBR) and connection-oriented routing. The service supported here may be, for instance, circuit emulation. Class B is similar but with variable bit rates (VBR) and it can support VBR video transmissions. Class C does not guarantee any time relationship; it supports VBR and connection-oriented routing. Finally, class D differs from class C by being connectionless. Currently three types of AAL are mainly employed.

	Class A	Class B	Class C	Class D
Time relationships	Yes		No	
Bit rate	Const	Variable		
Connection mode	Connection oriented			Connection less

Fig. 8.8   Classification of services

48 bytes

Fig. 8.9 Format of AAL1 SAR-PDU

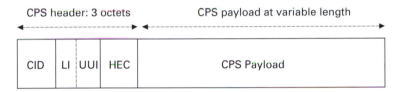

CPS header: 3 octets   CPS payload at variable length

Fig. 8.10 Format of AAL2-CPS

The AAL1 service, described in the document ITU-T I.363.1, supports class A applications. Constant bit rate services require data transfers at constant bit rates after the virtual circuit set-up between source and destination. To this end, AAL1 performs several functions, such as SDU transfer at constant bit rate, timing information transfer between source and destination, data structure information transfer or data loss indication. All these functions are important to support the circuit emulation service.

The SAR sublayer accepts from CS 47 octet blocks (the SAR-SDU) and adds one octet then forming the SAR-PDU (Fig. 8.9). At the destination, SAR removes the overhead octet and delivers the 47 octets to the upper CS. Considering the SAR header, the CS indication bit signals the presence of CS. The SN is the sequence number value provided by CS. Sequence number protection (SNP) is a four-bit field, which provides error detection and protection capabilities over the SAR-PDU header. Blocking of user data into 47 octet groups for SAR-SDU is performed at the CS sublayer. The AAL1 CS sublayer is service dependent and may perform several functions, such as handling cell delay variation to provide an AAL user with constant bit rate, handling SAR-PDU assembly delay, timing information transfers to recover at the destination the source clock frequency or reporting events of lost and misinserted cells.

The ITU-T I.363.2 standard describes AAL2, which can support class B services. It provides data transfers at variable bit rate and it allows for multiplexing different user information streams over one AAL2 virtual channel. Supporting VBR sources, it is possible to have not completely full cells, so that more SAR functions than AAL1 are required. The convergence sublayer is composed of two parts, the service specific convergence sublayer (SSCS), which may be null, and the common part convergence sublayer (CPCS). The CPCS with SAR is called common part sublayer (CPS). The AAL2-CPS is given by three octets and a payload of up to 45 bytes (Fig. 8.10). The payload size is reported in the length indicator (LI) field. The user-to-user (UUI) field

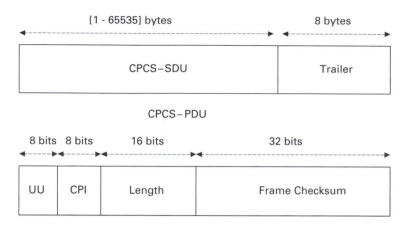

Fig. 8.11   Format of AAL5-CPCS

may be used by upper layers for data transfers. The eight-bit channel identifier (CID) field identifies a user data stream within an AAL2 virtual channel.

Document ITU-T I.363.5 describes AAL5, which is suitable for supporting service C class applications. The AAL5 layer is suitable for packet data transfers and it does not provide any guarantee to user data transfers. The CPCS-SDU can range from 1 to 65535 bytes in length and CPCS-PDU has a eight-byte trailer (Fig. 8.11). Then, it must be divided into groups to fit into cells. If its length, trailer included, is a multiple of 48 bytes, all cells are full, otherwise the last cell will have from 1 to 40 bytes, some padding and the trailer. At the receiver, cells will be reassembled into CPCS-PDU and the least meaningful bit of the payload type of the ATM header is used to mark the end of the CPCS-PDU.

8.6   MPLS principles

In the 1990s one of the goals of the scientific community working on networks was to increase the performance of IP routers by combining the IP routing with the ATM cell switching, i.e., routing IP datagrams over ATM. This means that the ATM switching matrix used for data forwarding is driven by IP routing rather than by the ATM control plane. This new router, also called *switch-router*, is supposed to operate as usual but, in certain cases, it can use ATM switching, avoiding hop-by-hop IP processing, by performing data cut-through, thus drastically reducing the latency time. This also means that it is possible to avoid reassembling and processing data units at the network level at each traversed node.

Two strategies were proposed for triggering data cut-through: one strategy was based on traffic data and the other on control information. In the former approach, called *IP switching*, ATM connections were set up in the presence of certain kinds of data traffic, properly recognized. In the latter, there were specific control protocols to establish, manage and tear down ATM connections. When these control protocols were based on

IP routing, the ATM virtual circuits reflected the IP routing topology and this approach was called *tag switching*. Whatever the approach is, the network architecture has to be able to define a mechanism for managing IP data flows in a differentiated way, and this goes beyond the pure connectionless paradigm adopted by IP.

This issue was so important that IETF established the working group for *multi-protocol label switching* (MPLS) with the goal of defining a reference solution. Multi-protocol label switching has been thought to combine robust and well-known IP routing with the flexibility of label switching, typical of ATM. With MPLS, routing tables are integrated by a *label information base* so that certain packet flows can be marked with a label. Once a label is assigned, it is then used for switching without IP processing anymore. In other words, a label is a fixed-length identifier that informs switch-routers about the proper way to manage the corresponding data unit and all units containing the same label value. Labels have to be managed, assigned and distributed to all nodes along a route and this is realized by means of a *label distribution protocol* (LDP), used by nodes to exchange label information [117].

A label-switched network (Fig. 8.12) is composed of several functional components: *label edge routers* (LERs), which are located at the borders and classify and add labels to data flows; *label switch-routers* (LSRs), which represent the core nodes and switch packets as a function of labels; LDP, as the protocol for label information exchange, which allows core routers to create the label information base and to keep it updated. Labels can be employed in different ways and inserted in diverse data units. For instance, to deploy MPLS over existing connection-oriented WANs, a label can be inserted in the VPI or VCI field of the ATM header or in the DLCI field of a frame relay frame, as shown in Fig. 8.13. As an alternative, label format and encapsulation used by MPLS to carry IP traffic over an Ethernet network is shown in Fig. 8.14.

When an LER realizes that a new packet flow must be label-switched, it forwards an LDP request to the following routers along the path toward the destination, to trigger the association between outgoing and incoming labels for that flow. When this request reaches the egress edge router, this router chooses a label value for identifying the flow

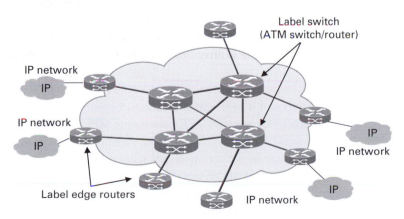

Fig. 8.12   Label-switched network

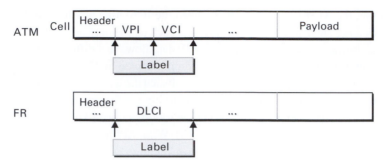

Fig. 8.13   MPLS label encapsulation in existing WAN environments

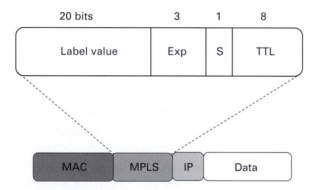

Fig. 8.14   MPLS label encapsulation in IP/LAN environments

on its incoming link and returns it to the upstream router along the path. The upstream LSR gets this label, stores it in its table for identifying the flow on its outgoing link, assigns another label to the same flow for identification on its incoming link, stores it in the table associated to the outgoing label and forwards it backward to the next upstream router. When a label is finally returned to the ingress edge router who sent the request, an edge-to-edge path is established and it is called a *label-switched path* (LSP). All the following packets of the same flow will then follow this path. One of the main results is then to provide connection-oriented functionalities to IP-based networks, as illustrated by the example in Fig. 8.15.

Recent advances in hardware-based processing and ultra-fast random access memories have boosted the evolution of IP router architectures to the point that the fast IP switching paradigm behind the original motivations for deploying MPLS networks is not critical anymore. However, network operators are still very keen on adopting MPLS in their core infrastructures because of the enormous advantages offered by such a technology that integrates the flexibility of the IP world with the traffic control and management capability achievable through a connection-oriented switching paradigm. In fact, today MPLS is mainly used to implement quality of service, traffic engineering and virtual private network solutions.

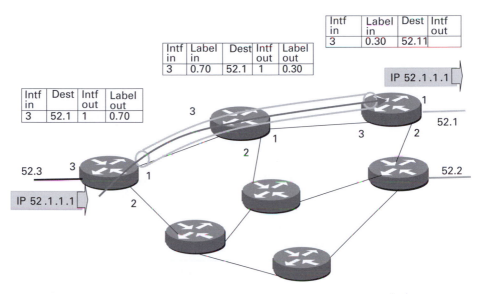

Fig. 8.15 Label-switched path

In recent years, MPLS has been extended to *generalized multi-protocol label switching* (GMPLS) [118], which is a multipurpose control plane paradigm able to manage not only packet switching devices, but also devices that perform switching in time, wavelength and space domains. Generalized multi-protocol label switching aims at extending the existing MPLS routing and signaling protocols to address the characteristics of optical transport networks. It defines a hierarchical LSP set-up. Label-switched paths may be tunneled inside an existing higher-order LSP, that acts as a link along the path of the lower-order LSP. Nodes at the border of two regions have the task of forming higher-order LSPs by aggregating lower-order LSPs.

8.7 Practice: MPLS router configuration

The purpose of this section is to provide a better understanding of the MPLS principles by illustrating a few practical cases of MPLS configuration on IP routers. In particular, the topics addressed include basic configuration of the label distribution protocol (LDP) and a simple case of an MPLS traffic engineering (MPLS-TE) solution. Owing to the peculiar nature of the topic and the limited availability of open-source MPLS implementations, the following examples refer to Cisco™ IP router configuration.

8.7.1 Basic LDP configuration

The first example illustrates the basic configuration of MPLS routers and the set up of LDP to distribute labels along the path between two LERs. The network configuration depicted in Fig. 8.16 shows two LERs interconnected by two LSRs representing the MPLS core network. Since MPLS operations rely on routing information at the network

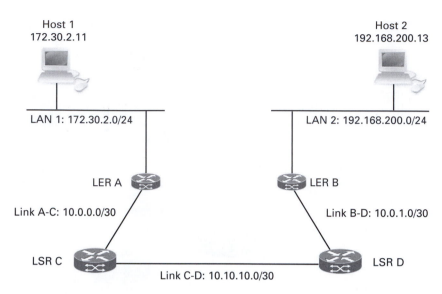

Fig. 8.16   MPLS network set-up with two LERs and two LSRs

layer, IP connectivity must be established before MPLS is activated. Therefore, the
router interfaces have been assigned IP addresses according to the address plan shown
in Fig. 8.16. Then, owing to the very simple topology used in this example, static routes
have been configured on each of the four routers. The commands used on LER A (in
configuration mode) are:

```
LER-A(config)#interface FastEthernet0/0
LER-A(config-if)#ip address 172.30.2.19 255.255.255.0
LER-A(config-if)#exit
LER-A(config)#interface FastEthernet0/1
LER-A(config-if)#ip address 10.0.0.2 255.255.255.252
LER-A(config-if)#exit
LER-A(config)#ip route 10.0.1.0 255.255.255.252 10.0.0.1
LER-A(config)#ip route 10.10.10.0 255.255.255.252 10.0.0.1
LER-A(config)#ip route 192.168.200.0 255.255.255.0 10.0.0.1
```

As shown by this example, the syntax of Cisco IOS™ operating system commands to
configure network interfaces and static routes recalls the `ip` command used in Linux
boxes, although there are some differences. For instance, an IP address can be assigned
to a router interface using the `ip address` command only after the desired interface
has been selected with the `interface` command. Static routes are added using the `ip
route` command. All the parameters required by these commands (i.e., prefix, netmask
and gateway) are the same as in a Linux box. The analogous configuration commands
for the other routers are:

```
LER-B(config)#interface FastEthernet0/0
LER-B(config-if)#ip address 192.168.200.19 255.255.255.0
LER-B(config-if)#exit
LER-B(config)#interface FastEthernet0/1
```

```
LER-B(config-if)#ip address 10.0.1.2 255.255.255.252
LER-B(config-if)#exit
LER-B(config)#ip route 10.0.0.0 255.255.255.252 10.0.1.1
LER-B(config)#ip route 10.10.10.0 255.255.255.252 10.0.1.1
LER-B(config)#ip route 172.30.2.0 255.255.255.0 10.0.1.1

LSR-C(config)#interface FastEthernet0/0
LSR-C(config-if)#ip address 10.0.0.1 255.255.255.252
LSR-C(config-if)#exit
LSR-C(config)#interface Serial0/2/0
LSR-C(config-if)#ip address 10.10.10.1 255.255.255.252
LSR-C(config-if)#exit
LSR-C(config)#ip route 10.0.1.0 255.255.255.252 10.10.10.2
LSR-C(config)#ip route 172.30.2.0 255.255.255.0 10.0.0.2
LSR-C(config)#ip route 192.168.200.0 255.255.255.0 10.10.10.2

LSR-D(config)#interface FastEthernet0/0
LSR-D(config-if)#ip address 10.0.1.1 255.255.255.252
LSR-D(config-if)#exit
LSR-D(config)#interface Serial0/2/0
LSR-D(config-if)#ip address 10.10.10.2 255.255.255.252
LSR-D(config-if)#exit
LSR-D(config)#ip route 10.0.0.0 255.255.255.252 10.10.10.1
LSR-D(config)#ip route 172.30.2.0 255.255.255.0 10.10.10.1
LSR-D(config)#ip route 192.168.200.0 255.255.255.0 10.0.1.2
```

Label distribution protocol operations require that each router is identified by a unique router-id, corresponding to the IP address assigned to one of the interfaces. Since a physical network interface is subject to changes in time – it could be shut down or its address may be modified – the best practice used to ensure protocol stability is to create a loopback interface, which is always active, and assign a fixed IP address to it. Then the LDP router-id will be configured based on this loopback interface. The network prefix that is added to a loopback interface is typically made by a single address, i.e., is a /32 prefix. This prefix must then be reachable by the other MPLS routers, so it must be included in the routing tables. For instance, in our example we want the loopback interfaces of LERs A and B to be assigned the addresses 10.9.9.1/32 and 10.9.9.2/32 respectively. The corresponding commands, including the required static routes, are:

```
LER-A(config)#interface Loopback1
LER-A(config-if)#ip address 10.9.9.1 255.255.255.255
LER-A(config-if)#exit
LER-A(config)#ip route 10.9.9.2 255.255.255.255 10.0.0.1

LER-B(config)#interface Loopback1
LER-B(config-if)#ip address 10.9.9.2 255.255.255.255
LSR-B(config-if)#exit
LER-B(config)#ip route 10.9.9.1 255.255.255.255 10.0.1.1
```

To configure LDP on a Cisco™ router, the MPLS forwarding function of IP packets must be enabled globally, as well as on each interface that will exchange MPLS packets. Then

LDP must be set as the protocol to be used for label distribution. Finally, the LDP router-id must be configured, specifying the loopback interface. The corresponding commands on LER A are:

```
LER-A(config)#mpls ip
LER-A(config)#interface FastEthernet0/1
LER-A(config-if)#mpls ip
LSR-A(config-if)#exit
LER-A(config)#mpls label protocol ldp
LER-A(config)#mpls ldp router-id Loopback1 force
```

Similar commands must be entered on the other routers. In the case of an LSR, the MPLS forwarding function of IP packets must be enabled on two interfaces, namely `FastEthernet0/0` and `Serial0/2/0`, according to the network configuration provided above.

Once LDP has been enabled, it is possible to verify which peering relationships have been established between MPLS routers. For instance, this can be done on LSR C with

```
LSR-C#show mpls ldp neighbor
      Peer LDP Ident: 10.9.9.1:0; Local LDP Ident 10.10.10.1:0
          TCP connection: 10.9.9.1.52067 - 10.10.10.1.646
          State: Oper; Msgs sent/rcvd: 1975/1972; Downstream
          Up time: 1d04h
          LDP discovery sources:
            FastEthernet0/0, Src IP addr: 10.0.0.2
          Addresses bound to peer LDP Ident:
            10.9.9.1      10.0.0.2
      Peer LDP Ident: 10.10.10.2:0; Local LDP Ident 10.10.10.1:0
          TCP connection: 10.10.10.2.39480 - 10.10.10.1.646
          State: Oper; Msgs sent/rcvd: 1968/1974; Downstream
          Up time: 1d04h
          LDP discovery sources:
            Serial0/2/0.1, Src IP addr: 10.10.10.2
          Addresses bound to peer LDP Ident:
            10.10.10.2
```

The output of the previous command shows the two LDP neighbors as seen by LSR C on the two interfaces using MPLS. The LDP router-id of LER A, i.e., 10.9.9.1, appears as the first peer, followed by the details about the connection established with it. Then, LSR D with router-id 10.10.10.2 is shown as the second peer. From the collected information it can be inferred that the LSRs have not been configured to use a loopback interface as LDP router-id.

While connectivity between LAN 1 and LAN 2 is still guaranteed by the IP routing tables, packet forwarding from router to router is now performed based on label switching instead of IP header look-up. This can be verified analyzing the traffic captures taken on the Ethernet link between LER A and LSR C and shown in Fig. 8.17.

When an ICMP echo request is sent from host 1 to host 2 (the upper box in the figure), LER A adds an MPLS header between Ethernet and IP headers and sets the Ethernet type field to the MPLS value (0x8847). In the example, traffic directed to LAN 2 is labeled

```
⊞ Frame 14 (102 bytes on wire, 102 bytes captured)
⊟ Ethernet II, Src: 00:1f:ca:0e:f9:f9, Dst: 00:1f:ca:0e:f9:80
    Destination: 00:1f:ca:0e:f9:80 (00:1f:ca:0e:f9:80)
    Source: 00:1f:ca:0e:f9:f9 (00:1f:ca:0e:f9:f9)
    Type: MPLS label switched packet (0x8847)
⊟ MultiProtocol Label Switching Header, Label: 19, Exp: 0, S: 1, TTL: 63
    MPLS Label: 19
    MPLS Experimental Bits: 0
    MPLS Bottom Of Label Stack: 1
    MPLS TTL: 63
⊞ Internet Protocol, Src Addr: 172.30.2.11 (172.30.2.11), Dst Addr: 192.168.200.13
⊟ Internet Control Message Protocol
    Type: 8 (Echo (ping) request)
    Code: 0
    Checksum: 0x89db (correct)
    Identifier: 0x7f0c
    Sequence number: 0x00c6
    Data (56 bytes)
⊞ Frame 15 (98 bytes on wire, 98 bytes captured)
⊟ Ethernet II, Src: 00:1f:ca:0e:f9:80, Dst: 00:1f:ca:0e:f9:f9
    Destination: 00:1f:ca:0e:f9:f9 (00:1f:ca:0e:f9:f9)
    Source: 00:1f:ca:0e:f9:80 (00:1f:ca:0e:f9:80)
    Type: IP (0x0800)
⊞ Internet Protocol, Src Addr: 192.168.200.13 (192.168.200.13), Dst Addr: 172.30.2
⊟ Internet Control Message Protocol
    Type: 0 (Echo (ping) reply)
    Code: 0
    Checksum: 0x91db (correct)
    Identifier: 0x7f0c
    Sequence number: 0x00c6
    Data (56 bytes)
```

Fig. 8.17   ICMP traffic capture on the link between LER A and LSR C

by LER A with value 19, as shown by the capture and by the following excerpt of the list of label bindings learned by LDP:

```
LER-A#show mpls ip binding 192.168.200.0 24
    192.168.200.0/24
        in label:      27
        out label:     19   lsr: 10.10.10.1:0     inuse
```

The previous command tells us that traffic directed to 192.168.200.0/24 has been assigned 27 as a local label value, while it is forwarded with label 19 to LSR C. In this case, only label 19 is actually used, since there are no upstream MPLS routers forwarding packets to that prefix via LER A. If there were, LER A (which is an LSR in this case) would notify them to use label 27, as the MPLS label value is always distributed by the downstream node to the upstream neighbor. The packet forwarded by LER A is then switched by the downstream LSRs, based on the label value and MPLS forwarding table, until it reaches its destination. In particular, when receiving packets labeled with 19, LSR C swaps the label to the value 32 and forwards the packet to LSR D, according to the third entry of its forwarding table:

```
LSR-C#show mpls forwarding-table
```

Local tag	Outgoing tag or VC	Prefix or Tunnel Id	Bytes tag switched	Outgoing interface	Next Hop
16	Pop tag	172.30.2.0/24	222068	Fa0/0	10.0.0.2
17	Pop tag	10.0.1.0/30	9856	Se0/2/0	point2point

```
19    32        192.168.200.0/24  398332   Se0/2/0   point2point
20    Pop tag   10.9.9.1/32       98253    Fa0/0     10.0.0.2
21    34        10.9.9.2/32       96997    Se0/2/0   point2point
```

The same procedure is applied in the opposite direction and, obviously, with different label values when an ICMP echo reply is sent back from host 2 to host 1. The capture of such a packet when it is forwarded from LSR C to LER A is shown in the bottom box of Fig. 8.17. It is interesting to notice that the captured packet does not include any MPLS label, even though it is forwarded from an MPLS node to another MPLS node. This is not a wrong behavior and it is perfectly in line with what is specified in the first entry of LSR C's forwarding table shown above, which tells the router to remove the MPLS header (*Pop tag*) before forwarding a packet to LER A when it is directed to LAN 1. The reason for this behavior is that packets forwarded from the second-last MPLS node (LSR C) to the last one (LER A) along the path are about to reach their final destination, as LER A is also in charge of delivering the packets directly to host 1. Since host 1 IP address look-up must be executed anyway by LER A, label processing is not required anymore and LSR C is allowed to send the packet without an MPLS label on the second-last network segment.

8.7.2   MPLS traffic engineering

The next example illustrates how to use MPLS to implement a simple solution of traffic engineering on the modified topology shown in Fig. 8.18, where a direct link connecting LERs A and B has been added. Now all the routers have been configured to use a dynamic routing protocol such as OSPF. Since the basic OSPF configuration on a Cisco™ router is very similar to what is done under Linux using Quagga, it is not shown here. In addition,

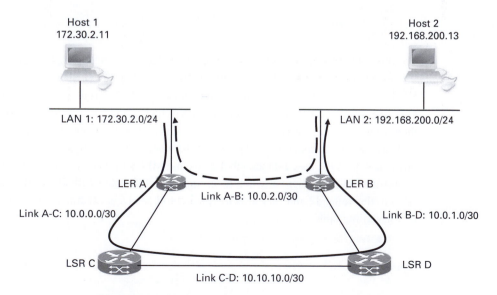

Fig. 8.18   Modified MPLS topology for traffic engineering

both LERs have been enabled to use MPLS also on the direct link. Therefore, all packets exchanged between the two LANs now use the shortest path represented by the link A-B. The objective is to use MPLS to force packets on a given direction to follow a different path specified by the operator, even though it is not the shortest one. More specifically, packets from LAN 1 to LAN 2 must follow the longest path (solid arrow), while packets in the opposite direction must keep following the shortest path (dashed arrow).

The MPLS packets will actually be forwarded by the routers through the desired paths because a suitable LSP will be established between the LERs. A possible way to achieve this on Cisco™ routers is to use the so-called MPLS traffic engineering (MPLS-TE) tunnels. The first thing to do is to enable MPLS-TE tunneling capabilities at the global router level, as well as on each interface traversed by the LSP. The corresponding commands to be executed on LSR-C are:

```
LSR-C(config)#mpls traffic-eng tunnels
LSR-C(config)#interface FastEthernet0/0
LSR-C(config-if)#mpls traffic-eng tunnels
LSR-C(config-if)#exit
LSR-C(config)#interface Serial0/2/0
LSR-C(config-if)#mpls traffic-eng tunnels
LSR-C(config-if)#exit
```

To establish MPLS-TE tunnels successfully, it is necessary to configure the underlying OSPF routing process to apply the traffic engineering (OSPF-TE) extensions. This requires the router-id to be specified and the OSPF area to be used as well as configuration of the tunnel characteristics in terms of allocated bandwidth:

```
LSR-C(config)#router ospf 100
LSR-C(config-router)#mpls traffic-eng router-id Loopback1
LSR-C(config-router)#mpls traffic-eng area 0
LSR-C(config-router)#exit
LSR-C(config)#interface FastEthernet0/0
LSR-C(config-if)#ip rsvp bandwidth 16
LSR-C(config-if)#exit
LSR-C(config)#interface Serial0/2/0
LSR-C(config-if)#ip rsvp bandwidth 16
LSR-C(config-if)#exit
```

In this example, the RSVP protocol will be used to reserve an amount of bandwidth equal to 16 kbps along the tunnel path. The value chosen is obtained as a quarter of the 64 kbps bandwidth available on the serial link between LSRs C and D. The previous commands must be executed on each MPLS router along the desired path.

Once the network has been prepared to use MPLS-TE tunnels, the ingress node (i.e., LER A) must be configured to establish a tunnel toward the egress node (i.e., LER B) following a specified path. The path chosen in the example, crossing LSRs C and D, can be configured by specifying the sequence of routers to be traversed as follows:

```
LER-A(config)#ip explicit-path identifier 10 enable
LER-A(cfg-ip-expl-path)#next-address 10.0.0.1
Explicit Path identifier 10:
    1: next-address 10.0.0.1
```

```
LER-A(cfg-ip-expl-path)#next-address 10.10.10.2
Explicit Path identifier 10:
    1: next-address 10.0.0.1
    2: next-address 10.10.10.2
LER-A(cfg-ip-expl-path)#next-address 10.0.1.2
Explicit Path identifier 10:
    1: next-address 10.0.0.1
    2: next-address 10.10.10.2
    3: next-address 10.0.1.2
LER-A(cfg-ip-expl-path)#exit
```

The final step is the definition of a tunnel interface, a sort of virtual interface without IP address that is typically associated to a loopback interface for stability purposes (remember that loopback interfaces are more stable than real interfaces). The other end of the tunnel, the MPLS-TE encapsulation mode and the explicit path previously defined to be used by the tunnel can finally be specified as follows:

```
LER-A(config)#interface Tunnel1
LER-A(config-if)#ip unnumbered Loopback1
LER-A(config-if)#tunnel destination 10.9.9.2
LER-A(config-if)#tunnel mode mpls traffic-eng
LER-A(config-if)#tunnel mpls traffic-eng path-option 1 expl id 10
LER-A(config-if)#exit
```

The tunnel is eventually established, as reported by the following excerpt of the show command (note the output label value set to 23 for packets forwarded to the tunnel):

```
LER-A#show mpls traffic-eng tunnels

Name: LER-A_t1 (Tunnel1) Destination: 10.9.9.2
  Status:
    Admin: up     Oper: up     Path: valid     Signalling: connected
    path option 1, type explicit 10 (Basis for Setup, path weight 66)

  Config Parameters:
    Bandwidth: 16 kbps (Global)  Priority: 7  7  Affinity: 0x0/0xFFFF
    Metric Type: TE (default)
    AutoRoute:  enabled   LockDown: disabled  Loadshare: 16  bw-based
    auto-bw: disabled

  InLabel  :  -
  OutLabel : FastEthernet0/1, 23
  RSVP Signalling Info:
      Src 10.9.9.1, Dst 10.9.9.2, Tun_Id 1, Tun_Instance 49
    RSVP Path Info:
      My Address: 10.0.0.2
      Explicit Route: 10.0.0.1 10.10.10.2 10.0.1.1 10.0.1.2
                      10.9.9.2
      Record Route:  NONE
      Tspec: ave rate=16 kbits, burst=1000 bytes, peak rate=16 kbits
[...]
```

Figure 8.19 shows ICMP traffic captured on the link between LER A and LSR C. Only ICMP echo requests are captured, with an MPLS label value set to 23 as expected. This

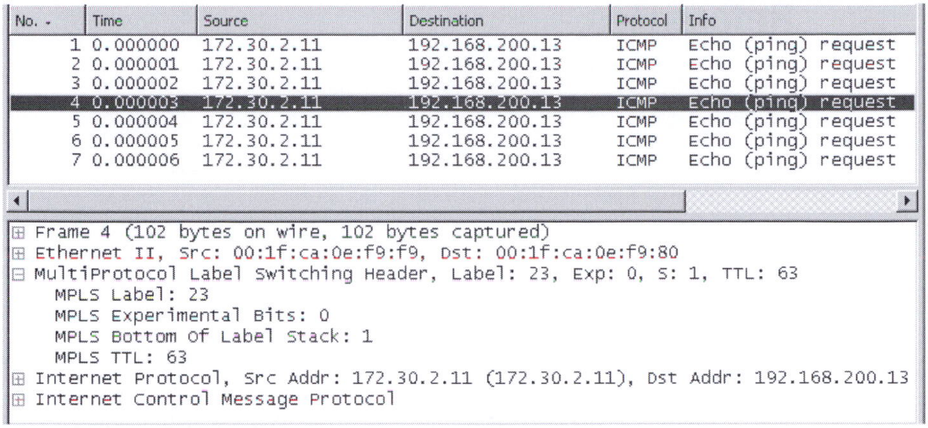

```
No. -   Time        Source          Destination       Protocol  Info
        1  0.000000  172.30.2.11     192.168.200.13    ICMP      Echo (ping) request
        2  0.000001  172.30.2.11     192.168.200.13    ICMP      Echo (ping) request
        3  0.000002  172.30.2.11     192.168.200.13    ICMP      Echo (ping) request
        4  0.000003  172.30.2.11     192.168.200.13    ICMP      Echo (ping) request
        5  0.000004  172.30.2.11     192.168.200.13    ICMP      Echo (ping) request
        6  0.000005  172.30.2.11     192.168.200.13    ICMP      Echo (ping) request
        7  0.000006  172.30.2.11     192.168.200.13    ICMP      Echo (ping) request

⊞ Frame 4 (102 bytes on wire, 102 bytes captured)
⊞ Ethernet II, Src: 00:1f:ca:0e:f9:f9, Dst: 00:1f:ca:0e:f9:80
⊟ MultiProtocol Label Switching Header, Label: 23, Exp: 0, S: 1, TTL: 63
    MPLS Label: 23
    MPLS Experimental Bits: 0
    MPLS Bottom Of Label Stack: 1
    MPLS TTL: 63
⊞ Internet Protocol, Src Addr: 172.30.2.11 (172.30.2.11), Dst Addr: 192.168.200.13
⊞ Internet Control Message Protocol
```

Fig. 8.19   ICMP traffic capture on the MPLS-TE tunnel between LER A and LER B

means that only the IP traffic from LAN 1 to LAN 2 is actually forwarded along the LSP established with the explicit non-shortest path, whereas the traffic in the opposite direction keeps following the shortest path.

The practical examples shown in this section were intended to give an insight of the steps that are typically required to configure MPLS routers and the available functionalities, such as traffic engineering.

8.8   Exercises

8.1   Q – How many subcarriers are employed in ADSL2+?

A – In ADSL2+512 subcarriers are employed operating in a frequency band up to 2.208 MHz.

8.2   Q – What kind of connections can X.25 establish?

A – X.25 is a connection-oriented network that supports two types of virtual circuit: permanent virtual circuit (PVC) and switched virtual circuit (SVC).

8.3   Q – What is the ISDN basic access?

A – The ISDN basic access is a user access given by two full-duplex B channels at 64 kbit/s and by one full-duplex D channel at 16 kbit/s.

8.4   Q – What is the core-edge approach adopted by frame relay?

A – Core-edge approach means that the processing functions of frame fields are divided into two groups: edge-to-edge and core functions. Several functions are done on edge-to-edge basis only, such as error recovery and flow control, not performed by core nodes.

8.5   Q – What is the basic ATM data unit?

A – It is a fixed size data unit composed of a 48 byte payload and a 5 byte header. The header includes the VPI/VCI field, which is used to manage the ATM virtual circuits.

References

[1] M. Schwartz, *Telecommunication Networks: Protocols, Modeling and Analysis*, Addison-Wesley, 1988.

[2] D. Bertsekas and R. Gallager, *Data Networks*, Prentice-Hall International, 1987.

[3] A. S. Tanenbaum, *Computer Networks*, Prentice-Hall International, 1996.

[4] J. K. Kurose and K. W. Ross, *Computer Networking*, Addison-Wesley, 1996.

[5] International Telecommunication Union, *ITU Homepage*, www.itu.int.

[6] European Telecommunications Standards Institute, *ETSI Homepage*, www.etsi.org.

[7] International Organization for Standardization, *ISO Homepage*, www.iso.org.

[8] Institute of Electrical and Electronics Engineers, *IEEE Homepage*, www.ieee.org.

[9] International Electrotechnical Commission, *IEC Homepage*, www.iec.ch.

[10] Internet Engineering Task Force, *IETF Homepage*, www.ietf.org.

[11] H. Alvestrand, *A Mission for the IETF RFC 3935*, Oct. 2004, Best Current Practice.

[12] International Standard ISO-IEC 7498-1, *Information Technology – Open Systems Interconnection, Basic Reference Model: The Basic Model*, 2nd edn., corrected and reprinted Jun. 1996.

[13] J. Postel, *Internet Control Message Protocol RFC 792*, Sept. 1981, Standard.

[14] ISI *Internet Protocol DARPA Internet Program Protocol Specification RFC 791*, Sept. 1981, Standard.

[15] R. Braden, ed., *Requirements for Internet Hosts – Communication Layers RFC 1122*, Oct. 1989, Standard.

[16] P. Almquist, *Type of Service in the Internet Protocol Suite RFC 1349*, Jul. 1992.

[17] K. Nichols, S. Blake, F. Baker and D. Black, *Definition of the Differentiated Services Field (DS Field) in the IPv4 and IPv6 Headers RFC 2474*, Dec. 1998, Proposed Standard.

[18] J. Postel, *Assigned Numbers RFC 820*, Aug. 1982, Historic.

[19] S. E. Deering, *Host Extensions for IP Multicasting RFC 1112*, August 1989, Standard.

[20] J. Reynolds and J. Postel, *Assigned Numbers RFC 1700*, Oct. 1994, Historic.

[21] Y. Rekhter, B. Moskowitz, D. Karrenberg, G. J. de Groot and E. Lear, *Address Allocation for Private Internets RFC 1918*, Feb. 1996, Best Current Practice.

[22] Internet Assigned Numbers Authority, *IANA Homepage*, http://www.iana.org.

[23] V. Fuller and T. Li, *Classless Inter-domain Routing (CIDR): The Internet Address Assignment and Aggregation Plan RFC 4632*, Aug. 2006, Best Current Practice.

[24] IANA, *IPv4 Address Space Registry*, www.iana.org/assignments/ipv4-address-space.

[25] W. Simpson, ed., *The Point-to-Point Protocol (PPP) RFC 1661*, Jul. 1994, Standard.

[26] A. Retana, R. White, V. Fuller and D. McPherson, *Using 31-Bit Prefixes on IPv4 Point-to-Point Links RFC 3021*, Dec. 2000, Proposed Standard.

[27] J. C. Mogul and J. Postel, *Internet Standard Subnetting Procedure RFC 950*, Aug. 1985, Standard.

[28] T. Pummill and B. Manning, *Variable Length Subnet Table For IPv4 RFC 1878*, Dec. 1995, Historic.

[29] J. Postel, *Transmission Control Protocol RFC 793*, Sept. 19981, Standard.

[30] M. Mathis, J. Mahdavi, S. Floyd and A. Romanow, *TCP Selective Acknowledgment Options RFC 2018*, Oct. 1996, Proposed Standard.

[31] J. Postel, *User Datagram Protocol RFC 768*, Aug. 1980, Standard.

[32] J. D. Case, M. Fedor, M. L. Schoffstall and J. Davin, *Simple Network Management Protocol (SNMP) RFC 1157*, May 1990, Historic.

[33] L. Kleinrock and F. Tobagi, Packet switching in radio channels: part I – carrier sense multiple-access modes and their throughput-delay characteristics, *IEEE Transactions on Communications*, vol. 23, no. 12, Dec. 1975, pp. 1400–1416.

[34] F. Tobagi, V. B. Hunt, Performance analysis of carrier sense multiple access with collision detection *Computer Networks*, vol. 5, no. 5, Nov. 1980, pp. 245–259.

[35] TIA, *EIA/TIA 568 (1991) Commercial Building Telecommunications Cabling Standard*, 1991.

[36] ISO, ISO/IEC 11801, *Information Technology – Generic Cabling for Customer Premises*, 2nd edn. 2002.

[37] BSI, *EN50173-1:2002 Information technology. Generic Cabling systems. General Requirements and Office Areas*, 2003.

[38] ANSI, TSB-67, *Transmission Performance Specifications for Field Testing of Unshielded Twisted-Pair Cabling Systems*, Oct. 1995.

[39] ANSI, *ANSI/TIA/EIA-568-B.2 Commercial Building Telecommunications Standard Part 2*, 2001.

[40] ANSI, TSB-95, *Additional Transmission Performance Guidelines for 4-Pair 100 Ohm Category 5 Cabling*, Oct. 1999.

[41] ANSI, *ANSI/TIA/EIA-568-B.2-1 Commercial Building Telecommunications Standard Part 2: Addendum 1*, 2002.

[42] ANSI, *TSB-155 Guidelines for the Assessment and Mitigation of Installed Category 6 Cabling to Support 10GBASE-T*, Mar. 2002.

[43] ISO, *ISO/IEC TR 24750:2007 Information Technology – Assessment and Mitigation of Installed Balanced Cabling Channels In Order To Support 10GBASE-T*, 2007.

[44] ANSI, *ANSI/TIA/EIA-568-B.2-10, Commercial Building Telecommunications Standard Part 2: Addendum 10*, Draft.

[45] D. Plummer, *Ethernet Address Resolution Protocol: Or Converting Network Protocol Addresses to 48.bit Ethernet Address for Transmission on Ethernet Hardware RFC 826*, Nov. 1982, Standard.

[46] J. Postel, *Multi-LAN Address Resolution RFC 925*, Oct. 1984.

[47] R. Finlayson, T. Mann, J. C. Mogul and M. Theimer, *A Reverse Address Resolution Protocol RFC 903*, Jun. 1984, Standard.

[48] R. Droms, *Dynamic Host Configuration Protocol*, Mar. 1997, Draft Standard.

[49] IEEE, *IEEE 802.11 WG, ANSI/IEEE Std 802.11, Information Technology—Telecommunications and Information Exchange Between Systems; Local and Metropolitan Area Networks; Specific Requirements, Part 11: Wireless LAN Medium Access Control (MAC) and Physical Layer (PHY) Specifications*, 1999.

[50] IEEE, *IEEE 802.11 WG, IEEE Std 802.11a-1999, Supplement to IEEE Standard for Information Technology Telecommunications and Information Exchange Between Systems; Local and Metropolitan Area Networks; Specific Requirements, Part 11: Wireless LAN Medium*

Access Control (MAC) and Physical Layer (PHY) Specifications: High-speed Physical Layer in the 5 GHz Band, 1999.

[51] IEEE, *IEEE 802.11 WG, IEEE Std 802.11b-1999, Supplement to IEEE Standard for Information Technology Telecommunications and Information Exchange Between Systems; Local and Metropolitan Area Networks; Specific Requirements, Part 11: Wireless LAN Medium Access Control (MAC) and Physical Layer (PHY) specifications: Higher-speed Physical Layer Extension in the 2.4 GHz Band*, 1999.

[52] IEEE, *IEEE 802.11 WG, IEEE Std 802.11g-2003, Supplement to IEEE Standard for Information Technology Telecommunications and Information Exchange Between Systems; Local and Metropolitan Area Networks; Specific Requirements, Part 11: Wireless LAN Medium Access Control (MAC) and Physical Layer (PHY) Specifications. Amendment 4: Further Higher Data Rate Extension in the 2.4 GHz Band*, 2003.

[53] IEEE, IEEE 802.11 WG, IEEE Std 802.11e-2005, *Supplement to IEEE Standard for Information Technology Telecommunications and Information Exchange Between Systems; Local and Metropolitan Area Networks; Specific Requirements, Part 11: Wireless LAN Medium Access Control (MAC) and Physical Layer (PHY) Specifications. Amendment 8: Medium Access Control (MAC) Quality of Service Enhancements*, 2005.

[54] IEEE, *IEEE 802.11 WG, IEEE Std 802.11i-2004, Part 11: Wireless LAN Medium Access Control (MAC) and Physical Layer (PHY) Specifications. Amendment 6: Medium Access Control (MAC) Security Enhancements*, 2004.

[55] G. Bianchi, I. Tinnirello and L. Scalia, Understanding 802.11e contention-based prioritization mechanisms and their coexistence with legacy 802.11 stations, *IEEE Network*, vol. 19, no. 4, Jul./Aug. 2005, pp. 28–34.

[56] Wi-Fi Alliance, *Wi-Fi Certified 802.11n Draft 2.0: Longer-Range, Faster-Throughput, Multimedia-Grade Wi-Fi Networks*.

[57] Wi-Fi Alliance, Homepage www.wi-fi.org/.

[58] IEEE, *IEEE Std 802.16-2004, Part 16 Air Interface for Fixed Broadband Wireless Access Systems*, 2004.

[59] IEEE, *IEEE Std 802.16e-2005, Part 16 Air Interface for Fixed and Mobile Broadband Wireless Access Systems*, 2005.

[60] Wi-MAX Forum, *Homepage*, www.wimaxforum.org.

[61] IEEE, *IEEE 802.15 WG, IEEE Std 802.15.3-2003, Part 15.3: Wireless Medium Access Control (MAC) and Physical Layer (PHY) Specifications for High Rate Wireless Personal Area Networks (WPANs)*, 2003.

[62] I. F. Akyldiz, X. Wang, A survey on wireless mesh networks, *IEEE Radio Communications*, Sept. 2005, pp. 23–30.

[63] Wireshark, *Homepage*, www.wireshark.org/.

[64] Aircrack Suite, *Homepage*, www.aircrack-ng.org/.

[65] IEEE, *IEEE 802.1Q-2005, IEEE Standard for Local and Metropolitan Area Networks Virtual Bridged Local Area Networks*, 2005.

[66] R. Perlman, *Interconnections: Bridges, Routers, Switches, and Internetworking Protocols*, Addison-Wesley, 2000.

[67] S. Keshav and R. Sharma, Issues and trends in router design, *IEEE Communications Magazine*, vol. 36, no. 5, May 1998, pp. 144–151.

[68] B. Carpenter and S. Brim, *Middleboxes: Taxonomy and Issues, IETF RFC 3234*, February 2002.

[69] W. R. Cheswick, S. M. Bellovin and A. D. Rubin, *Firewalls and Internet Security: Repelling the Wily Hacker*, 2nd edn., Addison-Wesley, 2003.

[70] K. Egevang and P. Francis, *The IP Network Address Translator (NAT) IETF RFC 1631*, May 1994. Informational.

[71] P. Srisuresh and M. Holdrege, *IP Network Address Translator (NAT) Terminology and Considerations, RFC 2663*, Aug. 1999. Informational.

[72] P. Srisuresh and K. Egevang, *Traditional IP Network Address Translator (Traditional NAT) RFC 3022*, Jan. 2001. Informational.

[73] J. Postel and J. Reynolds, *File Transfer Protocol (FTP), RFC 959*, Oct. 1985. Informational.

[74] S. Kent and R. Atkinson, *Security Architecture for the Internet Protocol, RFC 4301*, Dec. 2005, Proposed Standard.

[75] M. Holdrege and P. Srisuresh, *Protocol Complications with the IP Network Address Translator, RFC 3027*, Jan. 2001. Informational.

[76] J. Rosenberg, R. Mahy, P. Matthews and D. Wing, *Session Traversal Utilities for NAT (STUN) RFC 5389*, Oct. 2008, Proposed Standard.

[77] F. Audet and C. Jennings, *Network Address Translation (NAT) Behavioral Requirements for Unicast UDP IETF RFC 4787*, Jan. 2007, Best Current Practice.

[78] X. Fu, M. Stiemerling and H. Schulzrinne, ed., implications and control of middleboxes in the Internet, *IEEE Network*, vol. 22, no. 5, Sept./Oct. 2008.

[79] O. Andreasson, *Iptables Tutorial*, http://ipsysctl-tutorial.frozentux.net/ipsysctl-tutorial.html.

[80] B. Hubert, *Linux Advanced Routing and Traffic Control*, http://lartc.org/howto.

[81] C. Huitema, *Routing in the Internet*, 2nd edn., Prentice Hall, 2000.

[82] T. H. Cormen, C. E. Leiserson, R. L. Rivest and C. Stein, *Introduction to Algorithms*, 2nd edn., McGraw-Hill, 2001.

[83] G. Malkin, *RIP Version 2 RFC 2453*, Nov. 1998, Standard.

[84] J. Moy, *OSPF Version 2 RFC 2328*, April 1998, Standard.

[85] ISO, ISO/IEC 10589:2002(E), *Information Technology – Telecommunications and Information Exchange Between Systems – Intermediate System to Intemediate System Intra-Domain Routeing Information Exchange Protocol for Use in Conjunction with the Protocol for Providing the Connectionless-mode Network Service (ISO 8473)*, 2002.

[86] D. Oran, *OSI IS–IS Intra-domain Routing Protocol IETF RFC 1142*, Feb. 1990, Informational.

[87] R. Callon, *Use of OSI IS–IS for Routing in TCP/IP and Dual Environments RFC 1195*, Dec. 1990, Proposed Standard.

[88] Y. Rekhter and T. Li, *A Border Gateway Protocol 4 (BGP-4) RFC 4271*, Jan. 2006, Draft Standard.

[89] The XORP Team, *eXtensible Open Router Platform (XORP)*, www.xorp.org.

[90] K. Ishiguro *et al.*, *Quagga Routing Software Suite*, www.quagga.net.

[91] ITU, *ITU-T Recommendation G.995.1; Overview of Digital Subscriber Line (DSL) Recommendations*.

[92] ITU, *ITU-T Recommendation G.992.1; Asymmetric Digital Subscriber Line (ADSL) Transceivers*.

[93] ITU, *ITU-T Recommendation G.992.3; Asymmetric Digital Subscriber Line Transceivers 2 (ADSL2)*.

[94] ITU, *ITU-T Recommendation G.992.5; Asymmetric Digital Subscriber Line (ADSL) Transceivers Extended Bandwidth ADSL2 (ADSL2plus)*.

[95] ITU, *ITU-T Recommendation G.991.1; High Bit Rate Digital Subscriber Line (HDSL) Transceivers*.

[96] ITU, *ITU-T Recommendation G.991.2; Single-Pair High-Speed Digital Subscriber Line (SHDSL) Transceivers*.

[97]   ITU, *ITU-T Recommendation G.993.1; Very High Speed Digital Subscriber Line Transceivers (VDSL).*

[98]   ITU, *ITU-T Recommendation G.993.2; Very High Speed Digital Subscriber Line Transceivers 2 (VDSL2).*

[99]   ITU, *ITU-T Recommendation X.25; Interface Between Data Terminal Equipment (DTE) and Data Circuit-terminating Equipment (DCE) for Terminals Operating in the Packet Mode and Connected to Public Data Networks by Dedicated Circuit.*

[100] ITU, *ITU-T Recommendation X.21; Interface Between Data Terminal Equipment and Data Circuit-terminating Equipment for Synchronous Operation on Public Data Networks.*

[101] ITU, *ITU-T Recommendation X.21bis; Use on Public Data Networks of Data Terminal Equipment (DTE) Which is Designed for Interfacing to Synchronous V-Series Modems.*

[102] ITU, *ITU-T Recommendation X.32; Interface Between Data Terminal Equipment (DTE) and Data Circuit-Terminating Equipment (DCE) for Terminals Operating in the Packet Mode and Accessing a Packet-Switched Public Data Network Through a Public Switched Telephone Network or an Integrated Services Digital Network or a Circuit-Switched Public Data Network.*

[103] ITU, *ITU-T Recommendation V.36; Modems for Synchronous Data Transmission Using 60–108 kHz Group Band Circuits.*

[104] ITU, *ITU-T Recommendation V.37; Synchronous Data Transmission at a Data Signalling Rate Higher than 72 kbit/s using 60–108 kHz Group Band Circuits.*

[105] ITU, *ITU-T Recommendation V.38; A 48/56/64 kbit/s Data Circuit-terminating Equipment Standardized for use on Digital Point-to-Point Leased Circuits.*

[106] ITU, *ITU-T Recommendation I.324; ISDN Network Architecture.*

[107] ITU, *ITU-T Recommendation Q.921; ISDN User-Network Interface Data Link Layer Specification.*

[108] ITU, *ITU-T Recommendation Q.922; ISDN Data Link Layer Specification for Frame Mode Bearer Services.*

[109] ITU, *ITU-T Recommendation I.233; Frame Mode Bearer Services.*

[110] ITU, *ITU-T Recommendation Q.933; ISDN Digital Subscriber Signalling System No. 1 (DSS1) Signalling Specifications for Frame Mode Switched and Permanent Virtual Connection Control and Status Monitoring.*

[111] ITU, *ITU-T Recommendation Q.931; ISDN User-Network Interface Layer 3 Specification for Basic Call Control.*

[112] ITU, *ITU-T Recommendation I.370; Congestion Management for the ISDN Frame Relaying Bearer Service.*

[113] ITU, *ITU-T Recommendation I.326; Functional Architecture of Transport Networks Based on ATM.*

[114] ITU, *ITU-T Recommendation I.321; B-ISDN Protocol Reference Model and its Application.*

[115] ITU, *ITU-T Recommendation I.361; B-ISDN ATM Layer Specification.*

[116] ITU, *ITU-T Recommendation I.363; B-ISDN ATM Adaptation Layer Specification.*

[117] IETF, *IETF Request for Comments 3031; Multiprotocol Label Switching Architecture.*

[118] IETF, *IETF Request for Comments 3945; Generalized Multi-Protocol Label Switching (GMPLS) Architecture.*

Index

1000BASE-CX, 79
1000BASE-LX, 79
1000BASE-SX, 79
1000BASE-T, 79
100BASE-FX, 77
100BASE-T2, 77, 78
100BASE-T4, 77, 78
100BASE-TX, 77
10BASE-T, 75
4B/5B encoding scheme, 77
5-4-3 rule, 77
568 Commercial Building Telecommunication
        Standard, 80
802.11, 97
  ACK frames, 129
  authentication frame, 132
  beacon frame, 131
  deauthentication frame, 132
  probe request, 131
  RTS/CTS frames, 129
802.11 1999 edition, 98
802.11a, 98, 100
802.11b, 98, 99
802.11e, 98, 107
802.11g, 98, 100
802.11i, 98
802.11n draft, 112
802.16, 113
  best effort service, 123
  common part sublayer, 118
  convergence sublayer, 118
  downlink communication, 64
  FDD, 115
  MAC PDU, 122
  management connections, 121
  mesh topology, 114
  non-real-time polling service, 123
  point-to-multipoint, 114, 121, 123
  privacy sublayer, 118
  real-time polling service, 123
  TDD, 115
  transport connections, 121
  unsolicited grant service, 123
  uplink communication, 64

802.16e, 113
8B/10B encoding scheme, 79

access network, 8, 14
ACK, in TCP, 57
acknowledgment, 40, 74
acknowledgment number, in TCP header, 55
ACR, 86
ACRF, 86
ad-hoc network, 102
adaptive modulation, 114
address resolution protocol, see ARP
addressing, 40
ADSL, 38, 220
  discrete multi-tone modulation, 221
ADSL modem, 9
ADSL2, 221
ADSL2+, 221
AIFS, 108
Aloha, 62
American wire gage, see AWG
amplitude spectrum, 3
application layer, 33, 37
arbitration IFS, see AIFS
ARP, 38, 40, 88
  arp command, 89
  cache, 90
  table, 89
asymmetric digital subscriber line, see ADSL
asymmetric service, 19
asynchronous transfer mode, see ATM
ATM, 231
  adaptation layer, 234
  cell, 233
  classes of service, 234
  virtual channel, 231
  virtual path, 231
attenuation, 84, 85
attenuation to cross-talk ratio, see ACR
attenuation to cross-talk ratio far-end,
        see ACRF
auto-negotiation, 80
AWG, 84

B-ISDN, 231
back-off interval, 104
bandpass transmission, 4
bandwidth, 5
baseband transmission, 4
basic rate set, 107
BE, 123
beacon frame, 105
BER, 75
bidirectional service, 19
binary transmission, 1
bit error rate, *see* BER
bit rate, 1
bit time, 1
block acknowledgments, in 802.11n, 112
Bluetooth, 97, 98, 124
bridge, 137
    filtering bridge, 137
    learning process, 137
    plug-and-play, 137
    transparent bridges, 137
broadband ISDN, *see* B-ISDN
broadcast service, 19
broadcast transmission, 14, 15
burstiness factor, 18, 27, 28
bus topology, 12, 27

cable modem, 6
cable quality, 81
cabling standards, 80
cabling system, 61, 80, 81
carrier sense, 103
carrier sense multiple access, *see* CSMA
carrier sense strategies, 65
Category 3, 76
Category 5, 86
Category 5 UTP, 77
Category 5e, 81, 86
Category 6, 87
Category 6a, 87
Category 7, 76, 87
CBR, 18
CCK modulation, 100
CD, 66
CF-pollable stations, 107
CFP, 102, 105, 106
channel access, 32
channel access strategies, 61
checksum field, in TCP header, 55
checksum field, in UDP header, 58
CID, 118, 121
CIDR, 46, 50
CIDR notation, 50
circuit-switched network, 9
class A addresses, 43
class B addresses, 43

class D addresses, 43
class E, 87
class E addresses, 43
class F, 87
class Fa, 87
classful addressing scheme, 44, 46, 48
classless addressing scheme, 46, 47
classless inter-domain routing, *see* CIDR
clear-to-send, *see* CTS
coaxial cable, 5, 6, 75
collision, 62, 66, 71, 74, 103
collision detection, *see* CD
collision domain, 68, 69
collision domain, diameter, 76
collision domain, in fast Ethernet, 77
collision phenomenon, 62
communication channel, 4
congestion, 21
connection identifier, in 802.16, *see* CID
connection-oriented, 36
connectionless, 36, 40
connectionless MAC, 74
constant bit rate source, *see* CBR
contention period, *see* CP
contention-based strategies, 63
contention-free period, *see* CFP
core network, 9, 14
CP, 102
cross-talk, 84, 85
CSMA, 63
    one-persistent, 65
    non-persistent, 65
    p-persistent, 65
CSMA/CA, 102
CSMA/CD, 70
    one-persistent, 68, 70
CTS, 103

data encapsulation, 33
data field, 73
data-link layer, 32
datagram, 10, 32, 39
    network, 10
    technology, 10
DCF, 102–104
DCF IFS, *see* DIFS
delay, 14
delay-sensitive application, 19, 20
delivery traffic indication message, *see* DTIM
destination port number, 58
DHCP, 91
DIFS, 104–106
direct delivery, 52, 88
direct sequence spread spectrum, *see* DSSS
directed broadcast, 44, 46, 49
distributed coordinated function, *see* DCF

DNS, 92
dotted decimal notation, 42, 49
DSSS, 98, 99
DTIM, 105
duration/ID field, 103
dynamic control access schemes, 63

EDCA, 108, 109
   access category, 108
   back-off times, 108
   transmission opportunities, 108
   transmission opportunity, 108
EIA/TIA 568
   advantages, 82
   fault detection, 82
EIA/TIA 568-B.2 standard, 86
EIA/TIA TSB67, 84
EIA/TIA-568-B.2-1, 87
EIFS, 104
EN 50173, 81
endpoint, 54
end-to-end control, 33
ether, 7
Ethernet, 61, 68
   bus topology, 69
   MAC frame format, 61
   point-to-point connection, 70
   star topology, 69
ETSI, 30
European Telecommunications Standards Institute,
   see ETSI
extended IFS, *see* EIFS

fast Ethernet, 61
fast link pulses, 80
FCS, 74
FDD, 15, 64
FDMA, 62
FEXT, 85, 86
FIN, in TCP header, 55, 57
firewall, 168
   application-level gateway, 168
   packet filter, 168
   stateful, 168
   stateless, 168
fixed channel assignment, 62
flow control, 71, 79
FLP, 80
Fourier transform, 2
fragmentation, 40
frame aggregation, in 802.11n, 112
frame check sequence, 74
frame format, in Ethernet LAN, 72
frame relay, 227
   congestion control, 230
   traffic control, 229

frequency bandwidth, 3
frequency channel plan, 99, 100
frequency division duplex, *see* FDD
frequency domain, 1
frequency hopping, 98, 99
FTP, 39
full duplex, 70
full-duplex transmission, 15
full-duplex transmission, in Ethernet networks, 71

gateway, 39
generalized multi-protocol label switching, *see*
   GMPLS
gigabit Ethernet, 61
   NIC, 79
GMPLS, 239
guided propagation, 5

half duplex, 70
half-duplex transmission, 15
half-duplex transmission, in Ethernet networks, 70
hardware address, 72
HC, 108
HCCA, 108, 109
   transmission opportunities, 108
HCF, 108
HDSL, 221
header, 33
Hertz, 3
hidden terminals, 103
high bit rate digital subscriber line, *see* HDSL
high rate DSSS, 99
horizontal communication, 34
host-ID, 43, 49
host-to-network interface, 37
hub, 69, 137
hybrid coordination function, *see* HCF

IANA, 45
IBSS, 102
ICMP, 38, 39
IEC, 30
IEEE, 30
IEEE 802 project, 67
IEEE 802.11, 97
   frame format, 110
IEEE 802.11i, 133
IEEE 802.15, 124
IEEE 802.16, 36, 64
IEEE 802.1Q, 140
IEEE 802.3, 61, 67, 68
IEEE 802.3u, 77
IEEE 802.4, 67
IEEE 802.5, 67
   MAC sublayer, 67

IETF, 30, 37
ifconfig command, Linux, 93
IFS, 104, 106
impedance, 84, 86
independent basic set, *see* IBSS
indirect delivery, 52
infrared, 99
infrared region, 7
initial sequence number, *see* ISN
initial sequence number, in TCP header, 55
insertion loss, 85
integrated services digital network, *see* ISDN
interframe space, *see* IFS
inter-building backbone, 81
interface, 36
intermediate cross-connect, 81
International Electrotechnical Commission, *see* IEC
International Organization for Standardization, *see* ISO
International Telecommunication Union, *see* ITU
Internet, 8
    infrastructure, 8
Internet control message protocol, *see* ICMP
Internet Engineering Task force, *see* IETF
Internet layer, 37
Internet service provider, *see* ISP
intra-building backbone, 81
IP, 38, 40
    checksum, 42
    current version, 40
    header, 40
    identification field, 41
    options, 42
    padding, 42
    private addresses, 45
    TOS, 41
    total length field, 41
    TTL, 42
    type of service, in header field, 40
IP address, 43, 87
IP address translation, 45
IP alias, 95
IP packet delivery, 52
IP private addresses, 48
IP-TV, 109
IPv4, 40, 42
IPv6, 45
ISDN, 225
ISM band, 99
ISM frequency band, 98
ISN, 57
ISO, 30
ISO 7498, 31
ISO-OSI model, 31
ISO/IEC 11801, 81

ISO/IEC 24750, 87
ISP, 9, 46
ITU, 29

jitter, 21

LAN, 6, 11, 12, 30, 61
layer-1 bandwidth, 5
layer-2 bandwidth, 5, 20
layer-2 switch, 78
length field, in UDP header, 58
length/type, 73
limited broadcast, 44
Linux box, 89
local area network, *see* LAN
local loop, 6
loopback address, 44
loss-sensitive application, 19, 20

MAC, 32
    address, 72
    connection, 37
    correspondence between MAC and IP addresses, 87
MAC frame format, in Ethernet, 72
main cross-connect, 81
MAN, 12, 61
man-in-the-middle attacks, in LAN, 90
Manchester encoding, 75
maximum segment size, *see* MSS
medium access control, *see* MAC
mesh clients, 126
mesh network, 12
mesh routers, 126
mesh topology, 12, 27
metropolitan area network, *see* MAN
middlebox, 168
MIMO, in 802.11n, 112
minimum frame size, 73
modulation, 4
MPLS, 237
    label distribution protocol, 237
    label edge router, 237
    label information base, 237
    label switch-router, 237
    label switched path, 238
    router configuration, 239
    traffic engineering, 244
MSS, 55, 57
multi-protocol label switching, *see* MPLS
multi-home host, 94
multi-mode fiber, 7
multicast transmission, 15
multiple access, 61
multiple-input–multiple-output, *see* MIMO
multipoint service, 19

NAT, 169
   basic NAT, 169
   IP masquerading, 171
   NAPT, 171
   network address and port translation, 171
   PAT, 171
   port address translation, 171
   port forwarding, 172
NAV, 103
Net-ID, 43, 49
netmask, 49, 52, 92
network address translator, *see* NAT
network allocation vector, *see* NAV
network bandwidth, 5
network interface card, *see* NIC
network layer, 32
NEXT, 85
NIC, 68, 87, 92
   configuration, 92
non-CF-pollable, 107
nrtPS, 123

OFDM, 100
OFDMA, 114, 115
optical fiber, 5, 7, 75
options field, in TCP header, 55
organization unique identifier, *see* OUI
OSI
   interface, 35
   layer-1, 32
   layer-2, 32
   layer-3, 32
   layer-5, 33
   layer-6, 33
   layer-7, 33
   PCI, 35
   PDU, 35
   primitives, 35
   protocol, 35, 37
   SAP, 35
   SDU, 35
   service, 35
   terminology, 35
OSI Model, 31
OUI, 73

packet, 9, 10
packet delay, in a network, 15, 17, 19, 20
packet delay, on a link, 16
packet loss rate, 20
packet-switched network, 9, 10
pad field, 73
padding field, in TCP header, 56
pair-to-pair far-end cross-talk, *see* FEXT
pair-to-pair near-end cross-talk, *see* NEXT

PAN, 11
patch cord, 82
PC, 105
PCF, 102, 103
PCF IFS, *see* PIFS
peer communication, 34
personal area network, *see* PAN
phase spectrum, 3
physical address, 72
physical layer, 32
physical layer bandwidth, 5
piconet, 125
PIFS, 104, 105
ping, 22
point coordinated function, *see* PCF
point coordinator, *see* PC
point-to-point communication, 14
point-to-point link, 32, 47, 48
point-to-point protocol, *see* PPP
point-to-point service, 19
polling, 106
PPP, 48
preamble, 72
prefix, 46, 51
prefix length, 46
presentation layer, 33
priority classes, 21
processing delay, 15
propagation delay, 16
protocol, 36
protocol control information, *see* OSI PCI
protocol data unit, *see* OSI PDU
proxy ARP, 90
pseudoheader, 55
PSTN, 9, 30
public switched telephone network, *see* PSTN
PUSH, in TCP header, 55

QoS, 21, 41
QSTA, 108
QTag prefix, 74
quality of service, *see* QoS, 41
quality of service in WLAN, 107
queueing delay, 16

radio channel, 5, 7
random access category, 63
RARP, 91
regional Internet registries, *see* RIR
repeater, 69, 136
   multiport, 69
request for comments, *see* RFC
request-to-send, *see* RTS
reserved field, in TCP header, 55
reset, in TCP header, 55

retransmission, 75
retransmissions, after a collision, 71
reverse ARP, *see* RARP
RFC, 30
ring network, 12
ring topology, 12, 27
RIR, 46
RJ-45, 76
RJ45, 82
round-trip time, *see* RTT
route command, 93
router, 9, 161
    architecture, 164
    default route, 164
    forwarding information base, 166
    forwarding table, 163
    forwarding table look-up, 163
    functions, 163
    longest prefix match, 163
    Patricia trie, 167
    routing table, 163
    trie, 166
routing, 187
    autonomous system, 197
    centralized, 188
    deflection, 188
    deterministic, 187
    distributed, 188
    dynamic, 188
    flooding, 188
    graph representation, 188
    hierarchical, 196
    hot potato, 188
    probabilistic, 187
    routing area, 197
    shortest path, 189
    static, 187
routing algorithm, 187
    Bellman–Ford, 189
    Dijkstra, 192
routing protocol, 193
    BGP, 198
    border gateway protocol, *see* BGP
    comparison, 195
    count to infinity, 193
    distance vector, 193
    exterior, 197
    interior, 197
    intermediate system to intermediate system, *see* IS–IS
    IS–IS, 197
    link state, 194
    link state packet, *see* LSP
    LSP, 195
    open shortest path first, *see* OSPF
    OSPF, 200

path vector, 194
RIP, 198
routing information protocol, *see* RIP
split horizon, 194
triggered update, 194
routing tables, 51
rtPS, 123
RTS, 103
    frame, 102
RTT, 17, 20, 22, 23, 25

SACK option, 57
service, 36
service access point, *see* OSI SAP
service data unit, *see* OSI SDU
service level agreement, *see* SLA
session layer, 33
SHDSL, 222
short IFS, *see* SIFS
SIFS, 104, 106
signal analysis, 1
simple network management protocol, *see* SNMP
simplex transmission, 15
single-mode fiber, 7
single-pair high-speed digital subscriber line, *see* SHDSL
SLA, 20
SNMP, 58
socket, 54
source burstiness, 27
source port number, 58
spanning tree algorithm, 138
standards, 29
star topology, 12, 27
start of frame delimiter, 72
stateless operating mode, in routers, 11
static resource assignment, 62
station supporting QoS in 802.11e, *see* QSTA
statistical multiplexing, 10, 18
store-and-forward, 137
subnet, 48–50, 52
    address, 49
    field, 49
subnet-ID, 49
subnetting, 45, 48–50
    variable size, 50
superframe, 125
supernetting, 45, 50, 51
switch, 69, 78, 139
    cut-through, 139
    management, 151
    store-and-forward, 139
switch-router, 236
symmetric service, 19
SYN flag, 57
SYN, in TCP header, 55

tail, 33
TC, 81
TCP, 36, 38, 39, 54
    active close, 57
    active open, 57
    destination port number, 54
    header, 54
    ISN, in TCP header, 55
    passive close, 57
    passive open, 57
    pseudoheader, 55
    segment, 56
    sequence number, in TCP header, 55
    source port number, 54
TCP/IP
    protocol suite, 8, 37, 38, 40
TDD, 15
TDMA, 62
telecommunication closet, *see* TC
three-way handshake, 56
throughput, 20, 63, 66, 78, 97, 109, 118
TIA TSB-155, 87
TIA/EIA-568-B.2-Addendum 10, 87
time division duplex, *see* TDD
time domain, 1
time to live, *see* TTL
token, 67
traceroute, 25
transmission delay, 16
transmission types, 14
transport layer, 37
transport service provider, 33
transport service user, 33
tree topology, 12, 27
truncated binary exponential back-off,
        algorithm, 71
TSB95, 86
twisted pair, 5, 75

UDP, 36, 38, 39, 54, 58
    header, 58
    length, 58
UGS, 123
UHF, 7
ultra-high frequency, *see* UHF
UMTS, 7, 30, 38
unacknowledged MAC, 74
unicast transmission, 15
unidirectional service, 19
universal addressing scheme, 72
unshielded twisted pair, *see* UTP
URG, in TCP header, 55

urgent pointer field, in TCP header, 55
UTP, 6, 75

variable bit rate source, *see* VBR
VBR, 18
VDSL, VDSL2, 222
very high-speed digital subscriber line, *see* VDSL
virtual carrier sense, 102, 103
virtual circuit, 18, 32
    numbering, 26
    translation table, 26
virtual circuit network, 10
virtual local area networks, *see* VLAN
VLAN, 139
    benefits, 141
    dynamic, 142
    IEEE 802.1Q tagging, 147
    inter-VLAN connectivity, 149
    static, 141
    tagging, 146
VLAN tagging, 74
voice over IP, *see* VoIP
VoIP, 18

WAN, 12
WANs, 220
WEP, 111, 132
Wi-Fi Alliance, 113
Wi-Fi Multimedia, 113
Wi-Fi protected access, *see* WPA
WiMAX Forum, 114, 124
wide area networks, *see* WAN
window field, in TCP header, 55
wired bus, 14
wired equivalent privacy, *see* WEP
wireless LAN, *see* WLAN
wireless mesh networks, 125
Wireshark, 127
WLAN, 7, 12–14, 30, 32, 97
    5 GHz band, 100
    architecture, 101
    modulation schemes, 99, 100
    multirate support, 107
    output power, maximum allowable, 101
    QoS, 109
WMAN, 32, 36
WPA, 132
WPA2, 132, 133
WPA2-PSK, 133

X.25, 222
xDSL, 220